"An amazing—eye-opening, transformative, riveting—book from one of the greatest psychologists of our time."
—CAROL S. DWECK, PhD, AUTHOR OF *MINDSET*

Winner Books for a Better Life Award

THE MARSHMALLOW TEST

WHY SELF-CONTROL IS THE ENGINE OF SUCCESS

WALTER MISCHEL

NATIONAL BESTSELLER

Praise for Walter Mischel's

The Marshmallow Test

"Walter Mischel is one of the most influential psychologists of the twentieth century, and *The Marshmallow Test* will make him one of the most influential in this century, too. Thanks in no small part to his groundbreaking (and viral) findings, we know that self-control is among the most important contributors to human flourishing, and that though its biological roots run deep, it can be dramatically enhanced in individuals and societies. Here we have the maestro himself explain the science and its implications — wisely, accessibly, and humanely."

— Steven Pinker, Johnstone Professor of Psychology, Harvard University, author of *The Better Angels of Our Nature*

"The discoveries that grew out of the marshmallow studies add up to one of the most insightful research stories in the history of psychology. Whatever it is now, your view of human nature will change profoundly as you read this brilliant book."

— Daniel Kahneman, author of *Thinking, Fast and Slow*

"A fascinating book. It is such an addictive treat that I had no self-control in reading it, until I understood that I could actually improve my self-control, and from then on I was in marshmallow heaven. Stimulating, fun, clear, lively, and drawn from rigorous studies. It's not only accessible but very convincing. Seriously, I love it."

— Alan Alda, actor, writer, science communication advocate

"This masterwork is a profound and inspiring exploration of the essential question of how we struggle to regulate our own behavior and how we can more frequently win the battle for self-control."

— David Laibson, Department of Economics, Harvard University

"Mischel amply demonstrates precisely why self-control is such a powerful tool in life with a useful understanding of how it can be cultivated." — Sherryl Connelly, *New York Daily News*

"This is an amazing — eye-opening, transformative, riveting — book from one of the greatest psychologists of our time. Mischel delivers the powerful message that self-control can be enhanced, and shows us how!" — Carol S. Dweck, professor of psychology, Stanford University, author of *Mindset*

"Part memoir, part scientific analysis, and part self-help guide.... Mischel has traveled around the world to study delayed gratification in various cultural and socioeconomic contexts. The principles from the marshmallow test seemed to hold universally."
— Maria Konnikova, *The New Yorker*

"Walter Mischel's 'marshmallow test' — one now or two later if you can wait — with four-year-olds in the Bing Preschool at Stanford has become legendary, a harbinger of our understanding the lifetime role of cognitive control, self-discipline, and will. His book *The Marshmallow Test*, a charmingly told scientific story, makes clear the test is not just about youngsters, but is helpful to us all in the marshmallow moments we face through life. Mischel has written a wonderful book, engaging, enlightening, and profound." — Daniel Goleman, author of *Emotional Intelligence* and *Focus*

"Mischel uses his impressive experience along with others' related research in the field to explore the nature — and nurture — of willpower. He explains simply and elegantly the complex neural and cognitive components that affect our ability to self-regulate." — *Success*

"Fascinating." — Peggy Brown, *Newsday*

"This is the book we've all been waiting for. Parents, teachers, and policymakers have long yearned for the original architect of self-control research to tell us what to take away from the single most important experiment in social science history. In this masterpiece, Walter Mischel explains why waiting for two little treats at age four, rather than devouring one right away, is so prognostic of later life outcomes. Most important, he illustrates with solid research and insightful anecdote the most important claim of the book: that self-control can be taught and mastered."

— Angela Lee Duckworth, associate professor, Department of Psychology, University of Pennsylvania, and a 2013 MacArthur Foundation Fellow

"*The Marshmallow Test* is a tour de force. Despite its serious academic content, it wears its learning lightly.... It is to be hoped that this book will make Walter Mischel as much of a household name as his marshmallows are." — Natalie Gold, *The Times* (UK)

"A fascinating read.... Crisp, clear.... Mischel explains the latest research and helps readers understand better the surprising results of one of the best-known psychological experiments of all time. That alone is a considerable achievement — and makes the book well worth the wait." — *The Economist*

"What makes *The Marshmallow Test* so remarkable is not simply this great psychologist's ability to mine years of complex research to provide simple strategies anyone can use.... More, it is Mischel's compassion and commitment to making life better for individuals and society that shine through."

— Deborah Hopkinson, *BookPage*

"Mischel's work lies at the point where clinical psychology and self-improvement overlap.... His insights are fascinating and rewarding." — Nicholas Blincoe, *The Telegraph* (UK)

"Walter Mischel has changed psychologists' view of human potential, and *The Marshmallow Test* will change yours. The book is full of insights about self-control and how to master it, though it does create one impulse that is hard to resist — the desire to read the book cover to cover. It is both a fascinating story of a brilliant researcher at work and a recipe for how to change one's life."

— Timothy Wilson, Sherrell J. Aston Professor of Psychology, University of Virginia, author of *Redirect*

"This marvelous book is unique, and beautifully written from beginning to end. The range that Walter Mischel covers — from creative cognitive science to neuroscience to genetics — is breathtaking. This speaks for science at its best. Bravo!"

— Eric R. Kandel, MD, winner of the Nobel Prize in Physiology or Medicine; university professor, Department of Neuroscience, Columbia University; author of *The Age of Insight* and *In Search of Memory*

"The happy revelation of Mischel's book is that destiny is not determined by a swallowed or unswallowed marshmallow.... As simple as the marshmallow test is, this is a complex book that explores human nature, neuroscience, and genetics, enlivened by a sprinkling of anecdotes. It's also a book that can show you how to change your behavior: whether it's finally setting up that pension, cutting your alcohol intake, or shunning the marshmallows for good." — Rosamund Urwin, *Evening Standard*

"*The Marshmallow Test* is a wonderfully rich treat in itself, laden with advice and detailed research."

— Daisy Yuhas, *Scientific American*

THE MARSHMALLOW TEST

WHY SELF-CONTROL IS THE ENGINE OF SUCCESS

WALTER MISCHEL

LITTLE, BROWN AND COMPANY
New York Boston London

For Judy, Rebecca, Linda

Little, Brown and Company
Hachette Book Group
1290 Avenue of the Americas, New York, NY 10104
littlebrown.com

Originally published in hardcover as *The Marshmallow Test: Mastering Self-Control* by Little, Brown and Company, September 2014
First Little, Brown and Company trade paperback edition, September 2015

The illustration on page 62, "Mr. Clown Box," courtesy of Walter Mischel, published in W. Mischel, "Processes in Delay of Gratification," in *Advances in Experimental Social Psychology,* vol. 7, edited by L. Berkowitz (New York: Elsevier, 1974), 252. The illustration on page 125, "Future Self Continuity Scale," from H. Ersner-Hershfield and others, "Don't Stop Thinking About Tomorrow: Individual Differences in Future Self-Continuity Account for Saving," *Judgment and Decision Making* 4, no. 4 (2009): 280–286. Used with permission. The illustration on page 128, "Retirement Allocation, Current versus Retirement-Aged Self," is courtesy of Hal E. Hershfield; avatar created by Chinthaka Herath. The illustration on page 199, "*If-Then* Profiles for Jimmy and Anthony," was created from data in Y. Shoda, W. Mischel, and J. C. Wright, "Intraindividual Stability in the Organization and Patterning of Behavior: Incorporating Psychological Situations into the Idiographic Analysis of Personality," *Journal of Personality and Social Psychology* 67, no. 4 (1994): 674–687.

Excerpts from David G. Myers, "Self-Serving Bias," in *This Will Make You Smarter: New Scientific Concepts to Improve Your Thinking,* edited by John Brockman (New York: Doubleday, 2012), 37–38. Used with permission. Quotes from George Ramirez are printed with permission. Excerpts from *Sesame Street* script for episode 4412 are reprinted with permission. "Sesame Workshop"®, "Sesame Street"®, and associated characters, trademarks, and design elements are owned and licensed by Sesame Workshop.

ISBN 978-0-316-23087-2 (hc) / 978-0-316-33619-2 (int'l ed) / 978-0-316-23086-5 (pb)
Library of Congress Control Number: 2014018058

10 9 8 7 6 5 4 3 2 1

RRD-C

Printed in the United States of America

CONTENTS

THE MARSHMALLOW TEST

INTRODUCTION

As both my students and my children can testify, self-control does not come naturally to me. I have been known to call my students in the middle of the night to ask how the latest data analysis was going, though it began only that evening. At dinners with friends, to my embarrassment my plate is often the first to be clean, when others are far from done. My own impatience, and the discovery that self-control strategies can be learned, has kept me studying those strategies for a lifetime.

The basic idea that drove my work and motivated me to write this book was my belief, and the findings, that the ability to delay immediate gratification for the sake of future consequences is an acquirable cognitive skill. In studies initiated half a century ago, and still ongoing today, we've shown that this skill set is visible and measurable early in life and has profound long-term consequences for people's welfare and mental and physical health over the life span. Most important, and exciting for its educational and child-rearing implications, it is a skill open to modification, and it can be enhanced through specific cognitive strategies that have now been identified.

The Marshmallow Test and the experiments that have followed over the last fifty years have helped stimulate a remarkable wave of research on self-control, with a fivefold increase in the number of scientific publications just within the first decade of this century. In this book I tell the story of this research, how it is illuminating the mechanisms that enable self-control, and how these mechanisms can be harnessed constructively in everyday life.

It began in the 1960s with preschoolers at Stanford University's Bing Nursery School, in a simple study that challenged them with a tough dilemma. My students and I gave the children a choice between one reward (for example, a marshmallow) that they could have immediately, and a larger reward (two marshmallows) for which they would have to wait, alone, for up to 20 minutes. We let the children select the rewards they wanted most from an assortment that included marshmallows, cookies, little pretzels, mints, and so on. "Amy," for example, chose marshmallows. She sat alone at a table facing the one marshmallow that she could have immediately, as well as the two marshmallows that she could have if she waited. Next to the treats was a desk bell she could ring at any time to call back the researcher and eat the one marshmallow. Or she could wait for the researcher to return, and if Amy hadn't left her chair or started to eat the marshmallow, she could have both. The struggles we observed as these children tried to restrain themselves from ringing the bell could bring tears to your eyes, have you applauding their creativeness and cheering them on, and give you fresh hope for the potential of even young children to resist temptation and persevere for their delayed rewards.

What the preschoolers did as they tried to keep waiting, and how they did or didn't manage to delay gratification, unexpectedly turned out to predict much about their future lives. The more seconds they waited at age four or five, the higher their SAT scores and the better their rated social and cognitive functioning in adolescence. At age 27–32, those who had waited longer during the Marshmallow Test in preschool had a lower body mass index and a better sense of self-worth, pursued their goals more effectively, and coped more adaptively with frustration and stress. At midlife, those who could consistently wait ("high delay"), versus those who couldn't ("low delay"), were characterized by distinctively different brain scans in areas linked to addictions and obesity.

What does the Marshmallow Test really show? Is the ability to delay gratification prewired? How can it be taught? What is its downside? This book speaks to these questions, and the answers are often surprising. In *The Marshmallow Test*, I discuss what "willpower" is and what it is not, the conditions that undo it, the cognitive skills and motivations that enable it, and the consequences of having it and using it. I examine the implications of these findings for rethinking who we are; what we can be; how our minds work; how we can — and can't — control our impulses, emotions, and dispositions; how we can change; and how we can raise and educate our children.

Everybody is eager to know how willpower works, and everybody would like to have more of it, and with less effort, for themselves, their children, and their relatives puffing on cigarettes.

The ability to delay gratification and resist temptations has been a fundamental challenge since the dawn of civilization. It is central to the Genesis story of Adam and Eve's temptation in the Garden of Eden, and a subject of the ancient Greek philosophers, who named the weakness of the will *akrasia.* Over the millennia, willpower was considered an immutable trait — you either had it or you didn't — making those low in willpower victims of their biological and social histories and the forces of the momentary situation. Self-control is crucial for the successful pursuit of long-term goals. It is equally essential for developing the self-restraint and empathy needed to build caring and mutually supportive relationships. It can help people avoid becoming entrapped early in life, dropping out of school, becoming impervious to consequences, or getting stuck in jobs they hate. It is the "master aptitude" underlying emotional intelligence, essential for constructing a fulfilling life. And yet, despite its evident importance, it was excluded from serious scientific study until my students and I demystified the concept, created a method to study it, showed its critical role for adaptive functioning, and parsed the psychological processes that enable it.

Public attention to the Marshmallow Test increased early in this century and keeps escalating. In 2006, David Brooks devoted an editorial to it in the Sunday *New York Times*, and years later in an interview he conducted with President Obama, the president asked Brooks if he wanted to talk about marshmallows. The test was featured in *The New Yorker* in a 2009 Department of Science article, and the research is widely presented in television programs, magazines, and newspapers throughout the world. It is even guiding the efforts of *Sesame Street*'s Cookie Monster to master his impulse to voraciously devour cookies so that he may join the Cookie Connoisseurs Club. The marshmallow research is influencing the curriculum in many schools that teach a wide range of children, from those living in poverty to those attending elite private academies. International investment companies use it to encourage retirement planning. And a picture of a marshmallow has become an immediately understood opener to launch discussions of delay of gratification with almost any audience. In New York City, I see kids coming home from school wearing T-shirts that say *Don't Eat the Marshmallows* and large metal buttons declaring *I Passed the Marshmallow Test.* Fortunately, as the public interest in the topic of willpower increases, so does the amount and depth of scientific information on how delay of gratification and self-control are enabled, both psychologically and biologically.

In order to understand self-control and the ability to delay gratification, we need to grasp not only what enables it but also what undoes it. As in the parable of Adam and Eve, we see headline after headline that reveals the latest celebrity — a president, a governor, another governor, a revered judge and moral pillar of society, an international financial and political wizard, a sports hero, a film star — who blew it with a young intern, a housekeeper, or an illegal drug. These people are smart, and not just in their IQ intelligence but emotional and social intelligence as well — otherwise they could not have achieved their eminence. Then why do they act so stupid? And why do they have so much company in the many men and women who never make it into the headlines?

I draw on findings at the vanguard of science to try to make sense of this. At the heart of the story are two closely interacting systems within the human brain, one "hot" — emotional, reflexive, unconscious — and the other "cool" — cognitive, reflective, slower, and effortful. The ways in which these two systems interact in the face of strong temptations underlie how preschoolers deal with marshmallows and how willpower works, or doesn't. What I learned changed my long-held assumptions about who we are, the nature and expressions of character, and the possibilities for self-generated change.

Part I, Delay Ability: Enabling Self-Control, tells the story of the Marshmallow Test and the experiments that showed preschool children doing what Adam and Eve could not do in the Garden of Eden. The results identified the mental processes and strategies through which we can cool hot temptations, delay gratification, and achieve self-control. They also pointed to possible brain mechanisms that enable these achievements. Decades later, a flood of brain research is using cutting-edge imaging techniques to probe the mind-brain connections and help us understand what the preschooler managed to do.

The marshmallow findings inevitably lead to the question "Is self-control prewired?" Recent discoveries in the science of genetics are providing fresh answers to that question. They are revealing the surprising plasticity of our brains and transforming how we think about the role of nurture and DNA, environment and heredity, and the malleability of human nature. The implications go far beyond the science lab and contradict widely shared beliefs about who we are.

Part I leaves us with a mystery: why does the preschooler's ability to wait for more treats, rather than ring the bell and settle for less, predict so much about future success and well-being? I answer that question in Part II, From Marshmallows in Pre-K to Money in 401(k), where I look at how self-control ability influences the journey from preschool to retirement planning, how it paves the way to creating successful experiences and positive expectations — an "I think I can!" mind-set and a sense of self-worth. While not guaranteeing success and a rosy future, self-control ability greatly improves the chances, helping us make the tough choices and sustain the effort needed to reach our goals. How well it works depends not just on skills but on internalizing goals and values that direct the journey, and on motivation that is strong enough to overcome the setbacks along the route. How self-control can be harnessed to build such a life by making willpower less effortful and increasingly automatic and rewarding is the story of Part II, and like life itself it unfolds in unexpected ways. I discuss not just resistance to temptation but diverse other self-control challenges, from cooling painful emotions, overcoming heartbreak, and avoiding depression to making important decisions that take future consequences into account. And while Part II shows the benefits of self-control, it makes its limits equally clear: a life with too much of it can be as unfulfilling as one with too little.

In Part III, From Lab to Life, I look at the implications of the research for public policy, focusing on how recent educational interventions beginning in preschool are incorporating lessons on self-control in order to give those children living under conditions of toxic stress a chance to build better lives. I then summarize the concepts and strategies examined throughout this book that can help with everyday self-control struggles. The final chapter considers how findings about self-control, genetics, and brain plasticity change the conception of human nature, and the understanding of who we are and what we can be.

In writing *The Marshmallow Test*, I imagined myself having

a leisurely conversation with you, the reader, much like the many I have had with friends and new acquaintances, sparked by the question "What's the latest in the marshmallow work?" Soon we veer off into how the findings relate to aspects of our own lives, from child rearing, hiring new staff, and avoiding unwise business and personal decisions to overcoming heart-break, quitting smoking, controlling weight, reforming education, and understanding our own vulnerabilities and strengths. I have written the book for those of you who, like me, have struggled with self-control. I've also written it for those who simply would like to understand more deeply how our minds work. I hope *The Marshmallow Test* will start some new conversations for you.

PART I

DELAY ABILITY

Enabling Self-Control

PART I BEGINS IN the 1960s in what my students and I called "the Surprise Room" at Stanford University's Bing Nursery School, where we developed the method that became the Marshmallow Test. We started with experiments to observe when and how preschoolers became able to exert sufficient self-restraint to wait for two marshmallows they eagerly wanted rather than settle for just one right away. The longer we looked through the one-way observation window, the more we were astonished by what we saw as the children tried to control themselves and wait. Simple suggestions to think about the treats in different ways made it either impossibly difficult or remarkably easy for them to resist the temptation. Under some conditions they could keep on waiting; under others they rang the bell moments after the researcher left the room. We continued our studies to identify those conditions, to see what the children were thinking and doing that allowed them to control themselves, to try to figure out just how they made their struggles with self-control easier — or bound to fail.

It took many years, but gradually a model emerged of how the mind and brain work when preschoolers and adults struggle to resist temptations and manage to succeed. How self-control can be achieved — not by toughing it out or just saying "No!" but by changing how we think — is the story of Part I. Beginning early in life, some people are better than others at self-control, but almost everybody can find ways to make it easier. Part I shows how that can be done.

We also found that the roots of self-control are already visible in the toddler's behavior. So is self-control all prewired? Part I ends by answering that question in light of recent findings in genetics that profoundly change earlier views of the nature versus nurture puzzle. This new understanding has serious implications for how we raise and educate our children and how we think about them and ourselves, and I turn to this in subsequent chapters.

1

IN STANFORD UNIVERSITY'S SURPRISE ROOM

AT THE FAMOUS PARIS medical school named in honor of René Descartes, students crowd the street in front of its impressive pillared entry, chain-smoking cigarettes whose packets announce in French in large capital letters SMOKING KILLS. The messes that result when people cannot inhibit immediate gratification for the sake of delayed outcomes, even when they know they should, are familiar. We see them in our children and in ourselves. We see willpower's failure whenever earnest New Year's resolutions — to quit smoking, to go to the gym regularly, to stop quarreling with the person you love most — dissolve before January ends. I once had the pleasure of participating with Thomas Schelling, a Nobel laureate in economics, in a seminar on self-control. He wrote this summary of the dilemmas created by a weakness of will:

> How should we conceptualize this rational consumer whom all of us know and who some of us are, who in

self-disgust grinds his cigarettes down the disposal swearing that this time he means never again to risk orphaning his children with lung cancer and is on the street three hours later looking for a store that's still open to buy cigarettes; who eats a high-calorie lunch knowing that he will regret it, does regret it, cannot understand how he lost control, resolves to compensate with a low-calorie dinner, eats a high-calorie dinner knowing he will regret it, and does regret it; who sits glued to the TV knowing that again tomorrow he'll wake early in a cold sweat unprepared for that morning meeting on which so much of his career depends; who spoils the trip to Disneyland by losing his temper when his children do what he knew they were going to do when he resolved not to lose his temper when they did it?

Debates about the nature and existence of willpower notwithstanding, people go right on exercising it, struggling to climb up Mount Everest, enduring years of self-denial and strict training to get to the Olympics or star in the ballet, even kicking well-established drug addictions. Some adhere to stringent diets or give up tobacco after years of lighting the next cigarette from the one still in the mouth; others fail in spite of beginning with the same good intentions. And when we look closely at ourselves, how do we explain when and why our willpower and self-control efforts work or don't?

Before coming to Stanford as a psychology professor in 1962, I had done research on decision making in Trinidad and at Harvard, asking children to choose between less candy now or more later, or less money now versus more later. (I discuss this research in Chapter 6.) But our initial *choice* to delay and the ability to stick with it when faced with hot temptations easily go their separate ways. On entering a restaurant I can decide, indeed firmly resolve, "No dessert tonight! I won't do it because I have to avoid the cholesterol, the expanding waist, the next bad blood test..." Then the pastry cart rolls by and the waiter flashes the chocolate mousse in front of my eyes, and before there's time to reflect it winds up in my mouth. Given how often that happened to me, I became curious about what it takes to stick with the virtuous resolutions I kept abandoning. The Marshmallow Test became the tool for studying how people go from a choice to delay gratification to actually managing to wait and resist the temptation.

MAKING THE MARSHMALLOW TEST

From the age of antiquity, to the Enlightenment, to Freud, to the present day, young children have been characterized as impulsive, helpless, unable to delay gratification, and seeking only immediate satisfaction. With those naive expectations, I was surprised as I watched each of my three closely spaced daughters, Judith, Rebecca, and Linda, change in their first few years of life. They quickly morphed from mostly gurgling or screaming, to learning in exquisite detail how to annoy one another and enchant their parents, to becoming people with whom one could have fascinating, thoughtful conversations. In just a few years they could even sit more or less still to wait for things they wanted, and I tried to make sense out of what was unfolding in front of me at the kitchen table. I realized that I didn't have a clue about what went on in their heads that enabled them to control themselves, at least some of the time, and to delay gratification in the face of temptations, even when no one was hovering over them.

I wanted to understand willpower, and specifically delay of gratification for the sake of future consequences — how people experience and exert it, or don't, in everyday life. To move beyond speculation, we needed a method to study this ability in children as they began to develop it. I could see the skill developing in my three daughters when they were preschoolers at the Bing Nursery School at Stanford. This preschool was the ideal laboratory, newly completed on the campus as an integrated early education and research facility, with large one-way glass observation windows onto the attractive play areas, and small attached research rooms in which behavior could also be unobtrusively observed from a monitoring booth. We used one of these rooms for our research and told the children this was "the Surprise Room." That's where we escorted them to play the "games" that became our experiments.

In the Surprise Room, my graduate students Ebbe Ebbesen, Bert Moore, and Antonette Zeiss and I, as well as many other students, spent months of fun and frustration crafting, pilot-testing, and fine-tuning the procedure. For example, would telling preschoolers how long the delay would be — say 5 minutes versus 15 minutes — influence how long they waited? We found that it did not matter since they were still too young to understand such time differences. Would the relative amount of the rewards matter? It did. But what kind of rewards? We needed to create an intense conflict between an emotionally hot temptation that the child was eager to have immediately and one that was twice as large but required him or her to delay gratification for at least a few minutes. The temptation had to be meaningful and powerful enough for young girls and boys; appropriate, yet easily and precisely measurable.

Fifty years ago most children probably loved marshmallows as much as they do now, but — at least at Stanford's Bing Nursery School — their parents sometimes forbade them unless a toothbrush was at hand. Absent a universal favorite, we offered a selection of treats from which the children could choose. Whatever they selected, we offered them a choice of getting one treat right away or two if they waited for the researcher to return "by herself." Our frustration working out the details peaked when a first grant application to support the research was turned down by a federal agency with the suggestion that we apply instead to a candy company. We feared they might be right.

My previous research in the Caribbean had shown the importance of trust as a factor in the willingness to delay gratification. To assure that the children trusted the person who made the promise, they first played with the researcher until they were comfortable. Then the child was seated at a small table that had a desk bell on it. To further increase trust, the researcher repeatedly stepped out of the room, the child rang the bell, and the researcher immediately jumped back in, exclaiming, "You see? You brought me back!" As soon as the child understood that the researcher would always return immediately when summoned, the self-control test, described as another "game," began.

Though we kept the method simple, we gave it an impossibly cumbersome academic name: "The preschool self-imposed delay of immediate gratification for the sake of delayed but more valued rewards paradigm." Fortunately, decades later, after the columnist David Brooks discovered the work and featured it in the *New York Times* under the title "Marshmallows and Public

Policy," the media dubbed it "the Marshmallow Test." The name stuck, although we often did not use marshmallows as the treats.

When we designed the experiment in the 1960s we did not film the children. But twenty years later, to record the Marshmallow Test procedure and to illustrate the diverse strategies children use as they try to wait for their treats, my former postdoc Monica L. Rodriguez filmed five- to six-year-olds with a hidden camera in a public school in Chile. Monica followed the same procedure we had used in the original experiments. First up was "Inez," an adorable little first grader with a serious expression but a twinkle in her eye. Monica seated Inez at a small table in the school's barren research room. Inez had chosen Oreo cookies as her treats. On the table were a desk bell and a plastic tray the size of a dinner plate, with two cookies in one corner of the tray and one in the other corner. Both the immediate and the delayed rewards were left with the children, to increase their trust that the treats would materialize if they waited for them as well as to intensify their conflict. Nothing else was on the table, and no toys or interesting objects were available in the room to distract the children while they waited.

Inez was eager to get two cookies rather than just one when given the choice. She understood that Monica had to go out of the room to do some work but that she could call her back at any time by ringing the bell. Monica let Inez try ringing it a couple of times, to demonstrate that each time she rang Monica would immediately come back in the room. Monica then explained the contingency. If Inez waited for her to come back by herself, she got the two cookies. If she did not want to wait, she could ring the bell at any time. But if she rang the bell, or began to eat the treat, or left the chair, she'd get only the single cookie. To be sure that Inez understood the instructions fully, she was asked to repeat them.

When Monica exited, Inez suffered for an agonizing few moments with an increasingly sad face and visible discomfort until she seemed about to burst into tears. She then peeked down at the treats and stared hard at them for more than ten seconds, deep in thought. Suddenly her arm shot out toward the bell but just as her hand got to it, she stopped herself abruptly. Gingerly, tentatively, her index finger hovered above the bell's ringer, almost but not quite touching it, over and over, as if to tease herself. But then she jerked her head away from the tray and the bell, and burst out laughing, as if she had done something terribly funny, sticking her fist into her mouth to prevent herself from roaring aloud, her face beaming with a self-congratulatory smile. No audience has watched this video without oohing and laughing along with Inez in empathic delight. As soon as she stopped giggling, she repeated her teasing play with the bell, but now she alternately used her index finger to shush herself and stuck her hand in front of her carefully closed lips, whispering "No, no" as if to stop herself from doing what she had been about to do. After 20 minutes had passed, Monica returned "by herself," but instead of eating the treats right away, Inez marched off triumphantly with her two cookies in a bag because she wanted to take them home to show her mother what she had managed to do.

"Enrico," large for his age and dressed in a colorful T-shirt, with a handsome face topped by neatly cut blond bangs, waited patiently. He tipped his chair far back against the wall behind him, banging it nonstop, while staring up at the ceiling with a bored, resigned look, breathing hard, seemingly enjoying the loud crashing sounds he made. He kept banging until Monica returned, and he got his two cookies.

"Blanca" kept herself busy with a mimed silent conversation — like a Charlie Chaplin monologue — in which she seemed to be carefully instructing herself on what to do and what to avoid while waiting for her treats. She even mimed smelling the imagined goodies by pressing her empty hand against her nose.

"Javier," who had intense, penetrating eyes and an intelligent face, spent the waiting time completely absorbed in what appeared to be a cautious science experiment. Maintaining an expression of total concentration, he seemed to be testing how slowly he could manage to raise and move the bell without ringing it. He elevated it high above his head and, squinting at it intently, transported the bell as far away from himself as possible on the desktop, stretching the journey to make it as long and slow as he could. It was an awesome feat of psychomotor control and imagination from what looked like a budding scientist.

Monica gave the same instructions to "Roberto," a neatly dressed six-year-old with a beige school jacket, dark necktie on his white shirt, and perfectly combed hair. As soon as she left the room he cast a quick look at the door to be sure it was tightly shut. He then rapidly surveyed the cookie tray, licked his lips, and grabbed the closest treat. He cautiously opened the cookie to expose the white cream filling in its middle, and, with bent head and busy tongue, he began to lick the cream meticulously, pausing for only a second to smilingly approve his work. After licking the cookie clean, he skillfully put the two sides back together with even more obvious delight and carefully returned the filling-free cookie to the tray. He then hurried at top speed to give the remaining two cookies the identical treatment. After devouring their insides, Roberto arranged the remaining pieces on the tray to restore them to their exact original positions, and checked the scene around him, scanning the door to be sure that all was well. Like a skilled method actor, he then slowly sank his head to place his tilted chin and cheek on the open palm of his right hand, elbow resting on the desktop. He transformed his face into a look of utter innocence, his wide, trusting eyes staring expectantly at the door in childlike innocent wonder.

Roberto's performance invariably gets the most cheers and the loudest laughter and applause from every audience, including, once, a congratulatory shout from the esteemed provost of one of America's top private universities to "get him a scholarship when he's ready to come here!" I don't think he was joking.

PREDICTING THE FUTURE?

The Marshmallow Test was not designed as a "test." In fact, I have always had serious doubts about most psychological tests that try to predict important real-life behavior. I've often pointed to the limitations of many of the personality tests commonly used, and I've resolved never to create one myself. My students and I designed the procedure not to test children to see how well they did, but rather to examine what enabled them to delay gratification if and when they wanted to. I had no reason to expect that how long a preschooler waited for marshmallows or cookies would predict anything worth knowing about their later years, especially since attempts to predict long-term consequential life outcomes from psychological tests very early in life had been spectacularly unsuccessful.

However, several years after the marshmallow experiments began I started to suspect some connection between children's behavior in our experiments and how they fared later in life. My daughters had all attended the Bing school, and as the years passed I sometimes asked them how their friends from preschool were doing. Far from systematic follow-up, this was just idle dinnertime conversation: "How's Debbie?" "How's Sam doing?" By the time the kids were early teenagers, I started asking them to rate their friends on a scale of zero to five to indicate how well they were doing socially and in school, and I noticed what looked like a possible link between the preschoolers' results on the Marshmallow Test and my daughters' informal judgments about their progress. Comparing these ratings with the original data set, I saw a clear correlation emerging, and I realized that my students and I had to study this seriously.

It was 1978 and Philip K. Peake, now a senior professor at Smith College, was then my new graduate student at Stanford. Phil, working closely and often around the clock with other students, especially Antonette Zeiss and Bob Zeiss, was instrumental in designing, launching, and pursuing what became the Stanford longitudinal studies of delay of gratification. Beginning in 1982, our team sent out questionnaires to the reachable parents, teachers, and academic advisers of the preschoolers who had participated in the delay research. We asked about all sorts of behaviors and characteristics that might be relevant to impulse control, ranging from the children's ability to plan and think ahead, to their skills and effectiveness at coping with personal and social problems (for example, how well they got along with their peers), to their academic progress.

More than 550 children who were enrolled in Stanford University's Bing preschool between 1968 and 1974 were given the Marshmallow Test. We followed a sample of these participants and assessed them on diverse measures about once every decade after the original testing. In 2010, they reached their early to midforties, and in 2014, we are continuing to collect information from them, such as their occupational, marital, physical, financial, and mental health status. The findings surprised us from the start, and they still do.

ADOLESCENCE: COPING AND ACHIEVEMENT

In the first follow-up study, we mailed small bundles of questionnaires to their parents and asked them to "think about your child in comparison to his or her peers, such as classmates and other same-age friends. We would like to get your impression of how your son or daughter compares to those peers." They were to rate their children on a scale of 1 to 9 (from "Not at all" to "Moderately" to "Extremely"). We also obtained similar ratings from their teachers about the children's cognitive and social skills at school.

Preschoolers who delayed longer on the Marshmallow Test were rated a dozen years later as adolescents who exhibited more self-control in frustrating situations; yielded less to temptation; were less distractible when trying to concentrate; were more intelligent, self-reliant, and confident; and trusted their own judgment. When under stress they did not go to pieces as much as the low delayers did, and they were less likely to become rattled and disorganized or revert to immature behavior. Likewise, they thought ahead and planned more, and when motivated they were more able to pursue their goals. They were also more attentive and able to use and respond to reason, and they were less likely to be sidetracked by setbacks. In short, they managed to defy the widespread stereotype of the problematic, difficult adolescent, at least in the eyes and reports of their parents and teachers.

To measure the children's actual academic achievement, we asked parents to provide their children's SAT verbal and quantitative scores, when available. The SAT is the test in the United States that students routinely take as part of their application for college admission. To assess the reliability of the scores reported by the parents, we also contacted the Educational Testing Service, which administered the test. Preschoolers who delayed longer on the whole earned much better SAT scores. When the SAT scores of children with the shortest delay times (bottom third) were compared with those of children with longer delay times (top third), the overall difference in their scores was 210 points.

ADULTHOOD

Around age twenty-five to thirty, those who had delayed longer in preschool self-reported that they were more able to pursue and reach long-term goals, used risky drugs less, had reached higher educational levels, and had a significantly lower body mass index. They were also more resilient and adaptive in coping with interpersonal problems and better at maintaining close relationships (discussed in Chapter 12). As we continued to follow the participants over the years, the findings from the Bing study became more surprising in their sweep, stability, and importance: if behavior on this simple Marshmallow Test in preschool predicted (at statistically significant levels) so much for so long about how well lives turned out, the public policy and educational implications had to be considered. What were the critical skills that enabled such self-control? Could they be taught?

But perhaps what we were finding was a fluke, limited to what had been happening at Stanford, in the 1960s and early 1970s in California, at the height of the counterculture and the Vietnam War. In order to test this, my students and I launched a number of other studies with very different cohorts — not from the privileged Stanford campus community, but from very different populations and eras, including the public schools of the South Bronx in New York City decades after the Stanford studies had begun. And we found that things played out in similar ways with children living in extremely different settings and circumstances, which I describe in further detail in Chapter 12.

MIDLIFE BRAIN SCANS

Yuichi Shoda, now a professor at the University of Washington, and I have worked closely together since he started graduate school in psychology at Stanford in 1982. When, beginning in 2009, the Bing school participants reached their midforties, Yuichi and I organized a team of cognitive neuroscientists from several different institutions in the United States to conduct another follow-up study. This team included John Jonides at the University of Michigan, Ian Gotlib at Stanford, and BJ Casey at Weill Cornell Medical College. These colleagues were experts in social neuroscience, a field that focuses on understanding how the brain's mechanisms underpin what we think, feel, and do. They study these mechanisms with methods like functional magnetic resonance imaging (fMRI), which shows brain activity while an individual performs various mental tasks.

We wanted to test for possible differences in the brain scans of people whose lifelong trajectories, beginning with the Marshmallow Test, had been consistently either high or low on self-control measures. We invited a group of our Bing Nursery School alumni, who were now scattered in various parts of the country, to return for a few days to the Stanford campus, revisit the Bing school if they wanted, and take some cognitive tests, both while inside and outside the brain scanner at the Stanford School of Medicine, located on the same campus.

The brain images of these alumni revealed that those who had been more able to resist the marshmallow temptation in preschool and remained consistently high in self-control over the years displayed distinctively different activity in their frontostriatal brain circuitries — which integrate motivational and control processes — than those who hadn't. In the high delayers, the prefrontal cortex area, which is used for effective problem solving, creative thinking, and control of impulsive behavior, was more active. In contrast, in the low delayers, the ventral striatum was more active, especially when they were trying to control their reactions to emotionally hot, alluring stimuli. This area, located in the deeper, more primitive part of the brain, is linked to desire, pleasure, and addictions.

Discussing these findings with the press, BJ Casey noted that whereas low delayers seemed to be driven by a stronger engine, high delayers had better mental brakes. This study made a key point. Individuals who had lifelong low self-control on our measures did not have difficulty controlling their brains under most conditions of everyday life. Their distinctive impulse control problems in behavior and in their brain activity were evident only when they were faced with very attractive temptations.

2

HOW THEY DO IT

The Marshmallow Test and decades of subsequent studies showed us that self-control ability early in life is immensely important for how the rest of life plays out, and that this ability in the young child can be assessed at least roughly on a simple measure. The challenge was then to untangle the underlying mental and brain mechanisms that let some children wait for what seems like an unbearable amount of time during the test, while others ring the bell within seconds. If the conditions that facilitate self-control, and those that undermine it, could be identified, perhaps they could be harnessed to teach people who have trouble waiting to be better at it.

I chose preschoolers for the research because watching the changes in my own children suggested that this was the age at which youngsters begin to understand the contingency. They can grasp that if they choose the smaller treats now it prevents them from getting the more preferred treats later. It is also the age at which important individual differences in this ability become clearly visible.

DISTRACTION STRATEGIES

Many miracles seem to occur in the transformations from birth to crawling, talking, walking, and heading to preschool. No change was more remarkable to me than a child's transition from distressed howling for help to being able to wait, sitting alone in a chair with nothing to do, for many boring and frustrating minutes in anticipation of two cookies. How do they do it?

A century ago, Freud thought the newborn began as a completely impulse-driven creature, and he speculated about how this bundle of biological instincts that urgently pushed for immediate gratification managed to delay gratification when the maternal breast was withdrawn. In 1911, he proposed that this transition became possible in the first couple of years of life when the infant created a mental "hallucinatory image" of the objects of desire — the mother's breasts — and focused on it. In Freud's language, the infant's libidinal or sexual energy was directed at ("cathected onto") this hallucinatory image. This visual representation, he theorized, allowed "time binding"; it enabled the infant to delay and temporarily inhibit the impulse for immediate gratification.

The idea that mental representations of the reward and its anticipation sustained the goal-directed effort to pursue it was provocative — but it was not obvious how to test it with young children long before imaging machines could peek into the human brain. We figured that the most direct way to get the young child to mentally represent the anticipated rewards was to let her see them while she waited for them. In the first experiments, the child chose the rewards she wanted and then the researcher placed them on top of an opaque tray in front of her, in clear view. In other conditions, the researcher placed them right under the tray so they were covered and obscured from view. At this age, the children understood that their rewards were really there, underneath the tray. In what condition do you think it was hardest for the preschoolers to wait?

You probably intuitively guessed right: when the rewards were exposed, the temptation was great and it was hellish for the kids to wait; when the rewards were covered, it was easy. Preschoolers who were exposed to the rewards (whether the delayed ones, the immediate ones, or both) waited on average less than a minute, whereas they waited almost ten times longer when the rewards were covered. Although in retrospect the results seem obvious, we needed to demonstrate them to be sure we had found a truly tempting, difficult-conflict situation.

I watched the children unobtrusively through the one-way observation window while they were trying to wait in the rewards-exposed condition. Some covered their eyes with their hands, rested their heads on their arms to stare sideways, or turned their heads away to completely avoid facing the rewards. Trying desperately to avert their gaze for most of the time, some occasionally stole a quick glance toward the treats to remind themselves that they were still there and worth waiting for. Others talked quietly to themselves, their barely audible whispers seeming to reaffirm their intentions through self-instructions — "I'm waiting for the two cookies" — or by reiterating the choice contingency aloud: "If I ring the bell I'll get this one but if I wait I'll get those." Still others simply pushed the bell and the tray as far away from their faces and hands as they possibly could, right to the table's outer edge.

Successful delayers created all sorts of ways to distract themselves and to cool the conflict and stress they were experiencing. They transformed the aversive waiting situation by inventing imaginative, fun distractions that took the struggle out of willpower: they composed little songs ("This is such a pretty day, hooray"; "This is my home in Redwood City"), made funny and grotesque faces, picked their noses, cleaned their ear canals and toyed with what they discovered there, and created games with their hands and feet, playing their toes as if they were piano keys. When all other distractions were exhausted, some closed their eyes and tried to go to sleep — like one little girl who finally dropped her head into her folded arms on the table and fell into a deep slumber, her face inches from the signal bell. While these tactics were a marvel to behold in preschoolers, they are familiar to anyone who has ever been trapped in the front row at a boring lecture.

When going on long car trips with young children, parents often help their preschoolers generate their own fun to make the trip go faster. We tried that in the Surprise Room: before the waiting period began, we suggested that the children think some "fun thoughts" while waiting and prompted them to come up with a few examples, such as "when Mommy pushed me on a swing and I went up and down, all high up and down." Even the youngest children were wonderfully imaginative in generating their own fun thoughts when encouraged with a few simple examples. When happy thoughts were suggested before the researcher left the room, children waited for more than ten minutes on average, even when the rewards were exposed. Their self-generated fun thoughts counteracted the strong effects of exposure to the actual rewards, allowing them to wait as long as they did when the rewards were covered. They waited less than a minute when distracting thoughts were not primed. In contrast, cueing them to think about the rewards for which they were waiting (for example, "If you want to, while you're waiting you can think about the marshmallows") guaranteed that they would ring the bell soon after the door closed.

FROM DISTRACTION TO ABSTRACTION: "YOU CAN'T EAT A PICTURE"

To get the participants closer to forming the mental images Freud might have had in mind, we showed the children pictures of the treats rather than the treats themselves. Bert Moore, then my graduate student at Stanford (currently dean of the School of Behavioral and Brain Science at the University of Texas at Dallas), and I exposed preschoolers to realistic, life-size photos of the treats they had chosen. The images were displayed on the screen of a slide projector box (which was the best technology of the time) that was placed on the table at which the children sat. If the child had selected marshmallows, for example, then she saw a slide-projected image of them while she waited.

Now we got a big surprise: the results were completely reversed. Exposure to the real treats had made delay intolerable for most, but here we learned that exposure to their realistic images made it much easier to wait. Children who were exposed to images of the treats waited almost twice as long as those who saw irrelevant images or no images on the lit screen, or those who were exposed to the actual treats. Importantly, the images had to be of the treats for which the child was waiting, not of similar goodies that were irrelevant to what the child had chosen. In sum, an image of the object of desire, not the tempting object itself, made it easiest to wait. Why?

I asked "Lydia," a four-year-old girl with a smile-filled face, pink cheeks, and bright blue eyes, how she was able to wait the whole time, sitting patiently in front of the image of her treats. "You can't eat a picture!" she answered, as she happily began to sample her two marshmallows. When a four-year-old stares at the marshmallows she wants, she's likely to focus on their hot tempting features and ring the bell; when she sees a picture of them, it's more likely to serve as a cool reminder of what she'll get if she waits. As Lydia said, you can't eat a picture. And as Freud might have thought, you can't consume a hallucinatory representation of an object of desire.

In one condition of one of the studies, before the researcher exited, he said the following to children who were going to be looking at the real objects: "If you want to, when you want to, you can pretend they are not real, but just pictures; just put a frame around them in your head, like in a picture." Other children saw the picture of the rewards but were cued to think about them as if they were real: "In your head, you can make believe they're really there in front of you; just make believe they're there."

Facing pictures of the rewards, the children delayed 18 minutes on average — but when they pretended that the real rewards, rather than the pictures, were in front of them, they waited less than six minutes. Even when they faced the real rewards — the condition in which the average delay time is a minute or less — but imagined them as pictures, they could wait 18 minutes. The image they conjured up in their heads trumped what was exposed on the table.

HOT VERSUS COOL FOCUS

More than half a century ago, the Canadian cognitive psychologist Daniel Berlyne distinguished between two aspects of any stimulus. First, a tempting, appetitive stimulus has a consuming, arousing, motivating quality: it makes you want to eat the marshmallow, and when you do it's pleasurable. Second, it also provides descriptive cues that give information about its nonemotional, cognitive features: it's round, white, thick, soft, edible. So the effect the stimulus has on us depends on how we represent it mentally. An arousing representation focuses on the motivating, hot qualities of the stimulus — the chewy, sweet quality of the marshmallows or the feel of the inhaled cigarette smoke for the tobacco addict. This hot focus automatically triggers the impulsive reaction: to eat it or smoke it. A cool representation, in contrast, focuses on the more abstract, cognitive, informational aspects of the stimulus (it's round, white, soft, small) and tells you what it is like, without making it more tempting. It allows you to "think cool" about it rather than just grab it.

To test this idea, in one condition, before leaving the room, the researcher prompted the children to think about the hot, appetitive, appealing features of the rewards: the sweet, chewy taste of marshmallows. In a "think cool" condition, the children were prompted to think about the marshmallows as round and puffy clouds.

When cued to focus on the cool features of their rewards, children waited twice as long as when prompted to focus on the hot features. Importantly, when the child thought hot about the specific rewards for which he was waiting, it soon became impossible for him to continue to delay. But thinking hot about similar rewards for which he was not waiting (for example, pretzels while waiting for marshmallows) served as a splendid distraction and enabled an average of 17 minutes of delay. Children who just couldn't wait when cued to think "hot" about what they wanted right now could easily wait when cued to think "cool" about it.

The emotions the preschoolers experienced also affected how soon they rang the bell. If we suggested before leaving them alone with their temptations that while waiting they might think of some things that made them sad (like crying with no one to help them), they stopped waiting as fast as if we had suggested thinking about the treats. If they thought about fun things, they waited almost three times longer: close to 14 minutes on average. Give nine-year-old children compliments (for example, on their drawings), and they will choose delayed rather than immediate rewards much more often than when given negative feedback on their work. And what holds for children applies to adults. In short, we are less likely to delay gratification when we feel sad or bad. Compared with happier people, those who are chronically prone to negative emotions and depression also tend to prefer immediate but less desirable rewards over delayed, more valued rewards.

The hotter and more salient the desired reward, the more difficult it is to cool the impulsive reaction to it. Researchers offered almost seven thousand fourth and sixth graders in Israeli public schools choices between alternatives that varied in reward amounts (one versus two), delay time (immediate versus one week, one week versus one month), and appetitive appeal (chocolate, money, crayons). Not surprisingly, they chose the delayed alternatives most often for crayons and least often for chocolates. As every dieter knows, the hotness of a temptation exerts its power as soon as the refrigerator is open or the waiter describes the desserts.

The power is not in the stimulus, however, but in how it is mentally appraised: if you change how you think about it, its impact on what you feel and do changes. The tempting chocolate mousse on the restaurant dessert tray loses its allure if you imagine a cockroach just snacked on it in the kitchen. Although Shakespeare's Hamlet personified tragically unconstructive ways to appraise experience, he made this point insightfully: "There is nothing either good or bad, but thinking makes it so." As Hamlet also showed, trying to change how we think about or "mentally represent" stimuli and experiences that have become deeply ingrained can be as futile as trying to be your own brain surgeon. How one might cognitively reappraise events more easily and effectively is the central challenge for cognitive behavior therapies — and for anyone seriously committed to trying to change well-established dispositions and habits. It is also the basic question pursued throughout this book.

The marshmallow experiments convinced me that if people can change how they mentally represent a stimulus, they can exert self-control and escape from being victims of the hot stimuli that have come to control their behavior. They can transform hot tempting stimuli, and they can cool their impact by cognitive reappraisal — at least sometimes, under some conditions. The trick is getting the conditions right. It doesn't require Spartan clenched-teeth self-torture to toughen up and take the pain, but it does take more than strong motivation and the best intentions.

The power resides in the prefrontal cortex, which, if activated, allows almost endless ways of cooling hot, tempting stimuli by changing how they are appraised. The preschoolers, even with their immature frontal lobes, illustrated this with great imagination. They changed the temptations they faced into "just a picture" and put a frame around them in their heads; or shifted their attention away from temptations altogether through self-distraction, by inventing songs or exploring toes; or transformed them cognitively to focus on their cool and informative rather than hot and impulse-arousing features. When children transform marshmallows into puffy clouds floating in the air rather than thinking of them as delicious chewy treats, I have seen them sit in their chair with the treats and bell in front of them until my graduate students and I couldn't stand it anymore.

WHAT THE CHILDREN KNOW

We now knew that how children mentally represented external rewards predictably changed how long they waited. We also had learned in our other studies that children's ability to delay gratification increased with age, as did the range of strategies they could use to enable it. But what did the young child know about the strategies that would or would not be useful for helping him wait long enough to get those treats? How did the child's understanding of those strategies develop over time? Most important, did this understanding increase the ability to delay gratification?

My collaborators and I asked many children at different ages about the conditions, actions, and thoughts that would make it harder or easier for them to wait for their treats during the Marshmallow Test. None of these children had taken the test before, and they were introduced to it in the standard way. The child was seated at the little table, the selected treats were exposed on top of the tray, the bell was introduced, and the "one treat now or two later" contingency was explained. At this point, instead of leaving to let the child begin to wait, the researcher asked about the conditions that would help him or her wait. For example, "Would it be easier to wait if the marshmallows were on top of the tray so you could see them, *or* if they were under the tray so that you couldn't see them?"

At age three, most children could not understand the question and did not know what to say. Four-year-olds understood what we were asking but systematically selected the worst strategy: they wanted to expose the rewards during the delay period and to think about them, stare at them, and focus on how good they would be to eat. When asked why they were exposing the rewards, they said "Because it makes me feel good" or "I just want to see it" or "It's so yummy," apparently focusing on what they wanted ("I like them"), not yet understanding, or caring, that seeing the rewards would make it most difficult for them to wait. They wanted what they were waiting for to be right there in front of their eyes. And by having the rewards exposed they defeated their own solemn intentions to wait, surprising themselves when they saw that they had rung the bell and grabbed the treat. They not only failed to correctly predict their behavior, but they insisted on creating the conditions that would make it impossible for them to get the delayed rewards. These findings may help parents understand why their four-year-olds can still have such a hard time controlling themselves.

Within a year or so the change in the children was striking. By age five to six, most preferred to obscure the rewards and consistently rejected arousing thoughts about them as a strategy for self-control. Instead, they tried to distract themselves from the temptation ("Just sing a song" or "I guess I'll go to outer space" or "I think I'll take a bath"). As they got older, they also began to see the value of focusing on the contingency and reiterating it ("If I wait, I can get the two marshmallows, but if I ring, I'll get just one"). And they advised themselves with instructions: "I'll say, '*No*, do not ring the bell.' If I ring the bell and the teacher comes in, I'll just get that one."

"How should you wait for the marshmallows to make it easy?" I asked "Simon," age nine. He gave me his answer in a drawing of someone sitting during the Marshmallow Test, with a thought bubble showing that he was thinking about "something I like to distract myself." His additional written advice to me: "Don't look at what you are waiting for — don't think about nothing because then your [*sic*] thinking about it — Use what you have at the moment to entertain yourself." In further conversation, Simon explained how he managed this. He told me: "I have at least a thousand imaginary characters in my head, like those little toy figures I have in my room, and in my imagination I just take them out and play with them — I make up stories, adventures." Like Simon, other children his age can be wonderfully creative as they use their imaginations to entertain themselves and make the time pass quickly when they need to delay gratification in situations like the Marshmallow Test.

Most children did not seem to recognize the value of cool thoughts over arousing, hot thoughts until around age 12. By then, they usually understood that hot thoughts about the treats would defeat delay, whereas cool thoughts that transformed the marshmallows into puffy clouds, for example, would reduce their desirability and make it easier to wait. As one boy put it: "I can't eat puffy clouds."

The key question that drove this work was: does knowledge of the strategies that make delay easier also give the child — and the adult, for that matter — greater freedom from being controlled and pushed around by temptations and pressures they are trying to resist? We found the answer many years later in a study of boys with impulsivity problems who were living in a summer camp residential treatment program (described in Chapter 15). Those who understood strategies for delaying gratification waited longer on the Marshmallow Test than those who did not have this knowledge, and this was true even when the roles of age and verbal intelligence were controlled and removed statistically. It became clear that enhancing such understanding could become a goal for parents and teachers that might be fairly easy to achieve.

CAVEATS

In the 1980s I reported some of the early findings from the Stanford follow-up studies at a leading behavioral science research institute in Europe. I talked about the correlations we found between seconds of waiting on the Marshmallow Test and outcomes in adolescence, including SAT scores. A few months later, "Myra," a friend who was a senior researcher at the institute and had heard my talk, contacted me. She told me in all seriousness that she had some news that worried her. At age four, her son consistently refused to wait for more cookies (his favorites), no matter how hard she tried to get him to do so. An excellent scientist was misreading the meaning of the correlations I had reported. Myra was thinking, at least when it came to her son, that findings that were statistically significant and consistent for groups of children implied that if her child could not delay gratification on the measure she tried, it meant he faced a dire future.

When Myra calmed down, she of course realized how incorrectly she had interpreted the results: correlations that are meaningful, consistent, and significant statistically can allow broad generalizations for a population — but not necessarily confident predictions for an individual. Look at tobacco use, for example. Many people who smoke die early from tobacco-induced diseases. But some — indeed many — don't. If Johnny in preschool waits for his marshmallows you know that he is able to delay gratification, at least in that situation. If he doesn't, you can't be sure what it means. It could mean that he wanted to wait but couldn't, or simply that he had not used the bathroom before sitting down for the test. If a young child is eager to delay but finds herself ringing the bell, it's worth trying to understand the reasons.

As discussed in later chapters, some children start low in delay ability and get better at it over the years, and some start out eager and able to delay and then show decreasing levels of self-control over time. The experiments at the Bing Nursery School demonstrated how mental representations of temptations can change and even reverse their impact on behavior. The child who can't wait a minute can manage to wait for twenty when he changes his thoughts about the temptations. To me, that finding is more critical than the long-term correlations because it points the way to strategies that can enhance self-control ability and reduce stress. And advances in cognitive neuroscience and brain imaging in the last few decades have opened a window into the brain mechanisms underlying the ability to delay gratification. We can now begin to see how our thoughts can cool the brain when we most need to control our impulses.

3

THINKING HOT AND COOL

Once upon a time, by some estimates about 1.8 million years ago, our evolutionary ancestors were emerging from the trees of the river forest environment of the great apes. They were becoming *Homo erectus*, walking around on two feet in the grassy areas and struggling to live and reproduce. In these prehistoric adventures, the human species probably survived and multiplied thanks to the hot emotional system of the brain, the limbic system.

THE HOT EMOTIONAL SYSTEM

The limbic system consists of primitive brain structures located under the cortex on top of the brain stem, which developed early in our evolution. These structures regulate basic drives and emotions essential for survival, from fear and anger to hunger and sex. This system helped our ancestors cope with the hyenas, lions, and other wild beasts that were both their food supply and their daily mortal danger. Within the limbic system, the amygdala, a small almond-shaped structure (amygdala means "almond" in Latin), is especially important. It plays a key role in fear responses and in sexual and appetitive behavior. The amygdala rapidly mobilizes the body for action. It does not pause to think and reflect or worry about long-term consequences.

We still have a limbic system that works much as it did for our evolutionary ancestors. It remains our emotionally hot *Go!* system, specialized for quick reactions to strong, emotion-arousing stimuli that automatically trigger pleasure, pain, and fear. At birth it is already fully functional, making the infant cry when hungry or in pain. Although these days we rarely need it later in life for dealing with angry lions, it's still invaluable for avoiding menacing strangers in dark alleys or a swerving vehicle on an icy road. The hot system gives life its emotional zest. It motivates preschoolers to want two marshmallows, but it also makes it hard for them to endure the wait.

Activation of the hot system triggers instantaneous action: hunger for food and desire for other alluring stimuli elicit rapid hot *Go!* behaviors; threats and danger signals elicit fear and automatic defensive and flight reactions. The hot system is somewhat similar to what Freud called the id; he saw this as the unconscious structure of the mind, which contained sexual and aggressive biological impulses to seek immediate gratification and tension reduction, impervious to the consequences. Like Freud's id, the hot system operates automatically and mostly unconsciously, but it is in the service of much more than the sexual and aggressive impulses of Freud's concern. Reflexive, simple, and emotional, it automatically and quickly triggers consumptive behavior, arousal, and impulsive action. It makes the preschooler ring the bell and eat the marshmallow, the dieter bite into the pizza, the cigarette addict inhale the smoke, the angry abuser strike the partner, and the sexually out-of-control male grab the cleaning lady.

A focus on the hot features of a temptation easily triggers the *Go!* response. In the marshmallow experiments, I've watched a preschooler's hand suddenly lurch out and hit the bell hard, as the surprised child looks down in distress to see what his hand has done. For four-year-olds, the trigger can be anticipating the chewy, sweet taste of the marshmallows; for dieters, alcoholics, and smokers, each of the hot features has its own distinctive pull that can make its victims helpless. Even the sight or thought of the candy bar, or the whiskey, or the cigarette can elicit the action automatically. And the more often that happens, the more difficult it becomes to change the mental representation and avert the automatic *Go!* response. Learning and practicing some strategies for enabling self-control early in life is a lot easier than changing hot, self-destructive, automatic-response patterns established and ingrained over a lifetime.

High stress activates the hot system. This response was adaptive in evolutionary history for dealing with oncoming lions because it produces amazingly rapid (in milliseconds), automatic, self-protective reactions, and it is still useful in many emergencies in which survival requires instant action. But this hot response is not useful when success in a given situation depends on staying cool, planning ahead, and problem-solving rationally. And the hot system is predominant in the first few years of life, which makes it especially difficult for the young preschooler to exert self-control.

THE COOL COGNITIVE SYSTEM

Closely interconnected with the brain's hot system is its cool system, which is cognitive, complex, reflective, and slower to activate. It is centered primarily in the prefrontal cortex (PFC). This cool, controlled system is crucial for future-oriented decisions and self-control efforts of the kind identified in the Marshmallow Test. It's important to note that high stress attenuates the cool system and accentuates the hot system. The hot and cool systems continuously and seamlessly interact in a reciprocal relationship: as one becomes more active the other becomes less active. Although we rarely deal with lions, we daily face the endless stresses of the modern world, in which the hot system is often up, leaving us with our cool system down just when we need it most.

The PFC is the most evolved region of the brain. It enables and supports the highest-order cognitive abilities that make us distinctively human. It regulates our thoughts, actions, and emotions, is the source of creativity and imagination, and is crucial for inhibiting inappropriate actions that interfere with the pursuit of goals. It allows us to redirect our attention and to change strategies flexibly as the requirements of the situation shift. Self-control ability is rooted in the PFC.

The cool system develops slowly and becomes gradually more active in the preschool years and the first few years of elementary school. It does not fully mature until the early twenties, leaving the young child as well as the adolescent greatly vulnerable to the vicissitudes of the hot system. Unlike the hot system, the cool system is attuned to the informational aspects of stimuli and enables rational, reflective, and strategic behavior.

As I described in earlier pages, successful delayers in the Marshmallow Test invented ways to strategically distract themselves from the tempting treats and the bell. They also focused on the cool, abstract, informational features of the temptations as they imagined them (the marshmallows are like puffy clouds, or cotton balls), and avoided or transformed their hot features to cool them down (make believe it's just a picture; it's got a frame around it; you can't eat a picture). The diverse cognitive skills they used to wait for their treats are prototypes for those they needed years later to study for high school exams rather than heading out to the movies with friends, or countless other immediate temptations that awaited them in life.

Age matters. Most children younger than four are unable to sustain delay of gratification on the Marshmallow Test. When faced with the temptations, they ring the bell or start nibbling on the treats within about 30 seconds. Their cool system is not yet sufficiently developed. In contrast, by age 12 almost 60 percent of children in some studies have been able to wait even as long as 25 minutes, a very long time to be sitting facing a few cookies and a bell in a barren little room.

Gender also matters. Boys and girls develop different preferences at different phases of their development, and their willingness to wait will be influenced by the available rewards: what's rewarding to boys may be undesired by girls, and vice versa (fire engines, dolls, swords, makeup kits). But even if the reward values are equated and the motivation is the same, girls usually wait longer than boys, and their cooling strategies may differ. I have not measured it, but preschool boys seem to use more physical strategies, like tilting and rocking back and forth on the chair or pushing the temptations away, while girls seem to sing to themselves more or try to simply tune out. But that's my impression only, not a finding.

The greater willingness and ability of girls to wait longer is consistent with the finding that throughout the school years, at least in the United States, girls are usually rated higher on self-discipline measures than boys by their teachers, their parents, and themselves. Even in the first four years of life, girls are generally more compliant than boys. In later childhood, girls, on average, are usually seen as more self-disciplined in their schoolwork and they often get better grades than boys. However, the raters, including the children themselves, share cultural stereotypes about gender differences. "Good girls" are expected to be conscientious and careful, and "real boys" are supposed to be more impulsive, harder to control, and even rowdy, rehearsing their football tackles more than their times tables. On hypothetical choices about delayed rewards, like "Would you prefer $55 today or $75 in 61 days?," girls choose delayed rewards more often than boys. But when the choice becomes real, rather than hypothetical (keep an envelope containing a one-dollar bill today, or return it exactly one week later and get two dollars), the sex difference evaporates.

In short, we keep looking for sex differences on the Marshmallow Test and other measures of self-control. We don't always find them, but on the whole girls seem to have an advantage in the cognitive self-control skills and motivations that enable delay of gratification, at least in the populations and age groups studied so far.

When dealing with temptations, one way to momentarily escape the hot system is to imagine how someone else would behave. It's easier to use the cool system when making hot choices for others rather than for oneself. A researcher whose name I can't remember but whose story I can't forget asked preschoolers to consider a choice between a small piece of chocolate right now and a very large piece in ten minutes (he showed both pieces of chocolate to the children). When he asked a young boy, "What would an intelligent child choose?," the child responded that he would wait; when the researcher asked, "What will you do?," the child said, "I'll take it now!" The same point was made in an experiment with three-year-olds. They were given the choice between an immediate small reward and a delayed larger reward. When asked which one the experimenter would choose, they were able to use their cool system and were more likely to choose the delayed reward. But when they were choosing for themselves, the choice became hot and most of them took the smaller reward right away.

THE EFFECTS OF STRESS: LOSING THE COOL SYSTEM WHEN YOU NEED IT MOST

The experience of short-term stress can be adaptive, mobilizing you into action. Stress can become harmful, however, even toxic if it is intense and persists — for example, in people who become enraged at every frustration, from traffic jams to checkout lines, or who feel overwhelmed under extreme and enduring conditions of danger, turbulence, or poverty. Prolonged stress impairs the PFC, which is essential not only for waiting for marshmallows but also for things like surviving high school, holding down a job, pursuing an advanced degree, navigating office politics, avoiding depression, preserving relationships, and refraining from decisions that seem intuitively right but on closer examination are really stupid.

After reviewing research on the effects of stress, neuroscientist Amy Arnsten at Yale University concluded that "even quite mild acute uncontrollable stress can cause a rapid and dramatic loss of prefrontal cognitive abilities." The longer stress persists, the more those cognitive abilities are hurt and the more permanent the damage, ultimately leading to mental as well as physical illness. Thus, the part of the brain that enables creative problem solving becomes less available the more we need it. Remember Hamlet: as his stress escalated, he became increasingly trapped and tortured, paralyzed in his angry ruminations and fragmented feelings, and unable to think or act effectively, thus wreaking havoc all around him and further accelerating his undoing.

More than four hundred years after Shakespeare so eloquently dramatized Hamlet's mental anguish, we can reconstruct what must have been going on in his brain — not with the Bard's magical language, but with a model of the brain under chronic stress. The architecture of the brain is literally remodeled under chronic stress. Hamlet didn't stand a chance. When his stress persisted, his cool system, specifically the prefrontal cortex, crucial for problem solving, and the hippocampus, important for memory, began to atrophy. Concurrently, his amygdala, at the core of the hot system, increased excessively in size. This combination of brain changes made self-control and cool thinking impossible. Further, as his stress continued long term, his amygdala went from hypertrophy to atrophy, ultimately preventing normal emotional reactions. No wonder Hamlet was a tragedy.

4

THE ROOTS OF SELF-CONTROL

How early in life can you see the ability to delay gratification, or its lack, in your children? I discussed this often with friends when, like me, they were in the early child-raising years. Everybody was convinced that they saw the roots of such differences almost from birth. They were sure that Valerie had it and Jimmy didn't; Sam, yes, absolutely; Celia, not at all. It guaranteed vivid anecdotes and a lively conversation that left the question dangling, for future attention.

In 1983, about fifteen years after starting the marshmallow studies at Stanford, I accepted a professorship at Columbia University and moved back to New York City. One of the many enticements was that a young colleague, Lawrence Aber, was a faculty member at Barnard College, located across the street from Columbia. Larry was the research director of the Barnard Toddler Center, and we soon started a collaboration that continued for two decades. It was a chance to further study the dangling question about when and how delay ability develops.

THE "STRANGE SITUATION"

Waiting for marshmallows in the Bing Surprise Room may have been torture for four- and five-year-olds, but it was even tougher for an 18-month-old to have to wait for Mom to return after she had walked out of the little room at the Barnard Toddler Center and left her child alone with a stranger (a Barnard College volunteer) and some toys on the floor. Brief separations early in life are stresses that every child has to endure when the primary caretaker, usually the mother, begins to disappear, hopefully soon to return. By the middle of the second year of life, toddlers already differ greatly in how anxiously or securely or ambivalently attached they are to their primary caretakers. What they do during such separations and reunions allows a peek into the quality of their relationships and their coping skills early in life.

Mary Ainsworth designed the "Strange Situation" as a way to observe this relationship. Ainsworth was a student of John Bowlby, the highly influential British psychologist who, beginning in the 1930s, studied the effects of children's early-life attachment experiences, especially the impact of separation from their primary caregivers (an all-too-common, stressful experience in wartime). The Strange Situation simulates a brief maternal disappearance and reunion under controlled, benign conditions — Mom can come to the rescue quickly if her child's distress, expressed by heartbreaking crying or desperate pounding on the door, becomes excessive. The experiment plays out in three carefully scripted stages.

First, during *Free Play*, mother and child ("Benjamin" in this example) are left alone in the room together for five minutes to "play as you would at home."

Second, in the *Separation* episode, the school's director calls the mother out of the room, leaving Benjamin alone for 2 minutes with an undergraduate volunteer. Benjamin has previously seen or interacted with the volunteer in the presence of his mother for about 17 minutes. The volunteer remains silent during the separation unless Benjamin shows signs of distress, in which case she briefly reassures him that "mommies come back."

Third, during *Reunion*, right after the 2-minute separation, the mother reenters and picks Benjamin up. The volunteer leaves unobtrusively, and mother and child play together for 3 minutes.

In 1998, my student Anita Sethi wondered whether the 18-month-old's actions during separation predicted what he did three years later while waiting for his two marshmallows. To test this idea, we began at the Barnard Toddler Center where we set up the Strange Situation and video-recorded everything that happened during each stage. We noted the child's behavior during every ten-second frame — for example, whether he played or explored at a distance from the mother, distracted himself during her absence by looking at or playing with a toy, or engaged with the stranger. We also recorded his emotional expressions and any negative affect (crying, looking sad). The mother's spontaneous behavior was also coded in equal detail, including her attempts to initiate interaction with her child, her intrusions into his play or attempts to direct it, and her disregard of his cues. "Maternal control" — really *over*control and insensitivity to the child's needs — was rated, based on such cues as the mother's facial expression, vocal expression, position in relation to the child, amount of physical contact, expression of affection, and turn taking (sharing).

Toddlers who managed to distract themselves from Mom's absence by playing with the toys, exploring the room, or engaging with the stranger avoided the intense distress experienced by those who could not tear themselves away from the door and quickly dissolved into tears. The toddler's stress during the mother's 2-minute absence escalated with every second of delay. That last stretch of 30 seconds must have felt endless, and toddlers' behavior during those toughest seconds proved to be especially diagnostic: it predicted, far from perfectly but much beyond chance, what they would do when taking the Marshmallow Test in preschool. Specifically, the toddlers who spent those last 30 seconds of separation in the Strange Situation distracting themselves from Mom's absence became the ones who at age five waited longer for their treats and distracted themselves more effectively during the Marshmallow Test. In contrast, the toddlers who had been unable to activate the necessary distraction strategies were also unable to do so when waiting for their treats three years later and rang the bell sooner. These results underscore the importance of regulating attention to control and cool down stress, beginning early in life.

THE VULNERABLE ROOTS

At birth, infants are controlled almost completely by their internal state at each moment and by caregivers on whom they depend. In the infant's first few months outside the womb, soothing, rocking, feeding, and cuddling become a major job for caregivers, night and day. How lovingly and caringly infants are nurtured, or how cruelly and coldly they are neglected or abused, is inscribed in their brains and changes who they become. It is critical to keep infants' stress levels from becoming chronically activated and to promote the formation of close, warm attachments so the babies feel secure and safe.

The plasticity of the brain, especially in the first year of life, makes infants highly vulnerable to damage in their key neural systems if they have extremely adverse experiences, such as severe maltreatment or uncaring institutional rearing. Surprisingly, even much more moderate environmental stressors, such as exposure to persistent, albeit nonphysical, conflict between parents, may take a serious toll. In one study, while infants age 6 to 12 months were sleeping, their brains were scanned by fMRI. When they heard very angry-sounding speech while sleeping, the babies living with parents who had persistent conflicts, compared with those in less conflict-filled homes, had higher activation in the brain areas that regulate emotion and stress. Findings like these suggest that even relatively moderate stressors from the social environment during critical periods of development are registered in the hot system.

It is clear that as babies develop, their early emotional experiences are embedded into the architecture of their brains, and this can have huge consequences on how their lives unfold. Fortunately, interventions designed to enhance how babies regulate their emotions and develop cognitive, social, and emotional skills have the best chance to make a difference during those early years of life, when children are most vulnerable to damage. Within a few months of birth, caregivers can begin to switch their infants' attention away from feelings of distress and toward activities that interest them, and in time this helps their babies learn to self-distract to calm themselves. At the neural level, babies begin to develop the midfrontal area of the brain as an attention-control system for cooling and regulating their negative emotions. If all goes well, they become less reflexive, more reflective, less hot, more cool, and able to express their own goals, feelings, and intentions appropriately.

Discussing this process, Michael Posner and Mary Rothbart, two pioneers in the field of self-regulation development, say, "Children who at four months of age look at all the stimuli presented return to the lab a year and a half later with their own agenda. It is hard to get them to attend to our displays because their own plans take precedence. After making heroic efforts we can only shake our heads and mumble that *they have a mind of their own*" (emphasis mine).

As parents know, the second birthday is likely to come at around the time of the child's unwritten declaration of independence. In its early revolutionary phases, this striving for independence makes life challenging (to put it gently) for caregivers. Around the age of two to three years, children begin to be able to exert control over their thoughts, feelings, and actions, and this skill becomes increasingly visible during the fourth and fifth years of life. It is critical for success in the Marshmallow Test, as well as for adapting to school and beyond.

By age three, children can usually begin to make some purposeful choices, regulate their attention more flexibly, and inhibit impulses that would distract them from their goal. For example, studies by Stephanie Carlson and her colleagues at the University of Minnesota show that these children can manage to follow two simple rules — like "If it's blue, put it here, but if it's red, put it there" — long enough to reach their goal, often verbalizing self-instructions to help them figure out what they have to do. While impressive, these skills remain limited in the third year, but children make great strides in the following two years. By the time they reach their fifth birthday, their minds have become wonderfully sophisticated. There are large individual differences, of course, but many five-year-olds can understand and follow complex rules, like "If it's the color game, put the red square here, but if it's the shape game, put the red square there." While these skills are still in the early stages in the preschooler, by age seven, children's attention-control skills and the underlying neural circuits are surprisingly similar to those of adults. The child's experiences in the first half dozen years of life become roots for the ability to regulate impulses, exercise self-restraint, control the expression of emotions, and develop empathy, mindfulness, and conscience.

WHAT IF YOU HAVE A MOTHER LIKE PORTNOY HAD?

How does a mother's parenting style influence the self-control strategies and attachment that her child develops? In Anita Sethi's toddler study, described earlier, we measured the mother's behavior in detail to assess her level and style of "maternal control" and her sensitivity to her child's needs. Consider, for example, the mother who is overcontrolling and micromanaging, and who focuses mostly on her own needs rather than the child's. This profile is captured famously in Philip Roth's *Portnoy's Complaint.* As the protagonist looks back at his early childhood in New Jersey, he vividly recalls his mother's well-intended but stifling overcontrol: the intrusiveness with which she inspected, evaluated, and corrected everything from his arithmetic to the condition of his socks, nails, neck, and every crevice of his body. And when young Portnoy, overstuffed with his mother's loving cuisine, refuses to eat more pot roast, she persists, holding a long bread knife in her hand, and rhetorically asks: Does he want to become a skinny weakling, to be respected or ridiculed, "a man or a mouse"?

Portnoy's mother is a fictional creation, but I have friends who insist that their mothers were just like her. For a toddler who has Mrs. Portnoy for a mother, the path to achieving self-control skills may be very different from — indeed opposite — the one for the toddler with a less controlling mother. That is exactly what Anita found when she looked at the spontaneous interactions between the toddlers and their mothers when they were together in the room.

Toddlers who developed effective self-control skills by preschool generally responded to their highly controlling mother's bids for attention not by sticking close to her, but by distracting and distancing themselves from her (more than three feet) to explore the room and play with the toys. Toddlers who distanced themselves from their controlling mothers, who literally moved away when she made approach overtures, were able to delay longer on the Marshmallow Test at age five. They succeeded by using attention-control strategies to cool their frustration, distracting themselves from the rewards and the bell in the same way that, as toddlers, they had distracted themselves from their controlling mothers. In contrast, the toddlers who had equally controlling mothers but who hung close when Mom made her bids for attention focused on the temptations when they took the Marshmallow Test and quickly rang the bell.

For toddlers whose mothers were less controlling, it was a different story. When those mothers tried to engage their toddlers, the ones who stayed close were the ones who showed more effective self-control and cooling strategies during the Marshmallow Test at age five. They distracted themselves strategically, focused less on the temptations, and waited longer to get their bigger rewards than children who, as toddlers, had distanced themselves from such mothers.

What are the implications? A toddler with a mother who is not excessively controlling and is sensitive to his needs has no reason to distance himself from her, and he stays close when she approaches in the Strange Situation to reduce his stress. But what if a child has a mother who is highly sensitive to what she wants but blind to what her child needs when he most needs it, and who tries to control his every move in ways that distress him? Anita's results raise some questions to consider. It may not be a bad idea for the toddler to move a few feet away from his mom to play with the toys and explore the room. It may even help him develop the self-control cooling skills that he'll need to get those two marshmallows when he is age five.

To examine these possibilities, Annie Bernier of the University of Montreal led a research team in 2010 that studied how mothers interacted with their children, aged 12 to 15 months, to see how those interactions influenced the development of self-control. The researchers carefully examined how the mothers engaged with their toddlers when they worked together on puzzles and other cognitive tasks. Then they tested the same children again at age 16 to 26 months. Bernier found that the children of mothers who had in the earlier study encouraged their toddlers' autonomy by supporting their choices and sense of volition subsequently had the strongest cognitive and attention-control skills of the kind needed for success on the

Marshmallow Test. This was true even when the researchers controlled for differences in the mothers' cognitive ability and education. The message here is that parents who overcontrol their toddlers risk undermining the development of their children's self-control skills, while those who support and encourage autonomy in problem-solving efforts are likely to maximize their children's chances of coming home from preschool eager to tell them how they got their two marshmallows.

5

THE BEST-LAID PLANS

HOMER'S ANCIENT GREEK LEGEND *The Odyssey* tells of the adventures of Odysseus (Ulysses in the Roman version), the king of a small rugged island called Ithaca on Greece's western coast. The king leaves his new wife, Penelope, and their infant son at home and sets sail to fight in the Trojan War. Unexpectedly, the war drags on for many years — and so does Odysseus's journey home, which is filled with fantastic adventures of wild new loves, terrible battles, and tangles with horrific monsters. As he tries at long last to return home with his remaining sailors, they approach the land of the wondrous Sirens, whose irresistibly seductive voices and songs so enchant sailors on passing ships that they crash into the rocks and drown.

Odysseus was desperately eager to hear the Sirens' songs, but he was also aware of the hazards. In one of the Western world's earliest chronicles of advance planning done to resist temptation, he ordered his sailors to tie him firmly to the vessel's mainmast and leave him bound there — "and if I beseech and bid you to set me free, then do ye straiten me with yet more bonds." To protect themselves and assure that he remained tied, the sailors were ordered to plug up their ears with beeswax.

MR. CLOWN BOX

In the early 1970s, when the marshmallow experiments were well under way, I vaguely remembered Homer's tales. I also wondered whether Adam and Eve might have held on to paradise longer if they had had plans at the ready to help them resist snakes and apple temptations. I started to think about the Bing preschoolers: how would they deal with a powerful tempter who seduced their attention while they struggled to avoid the heavy costs of succumbing to him? Would advance planning help them resist? At that time, Charlotte Patterson, now a professor at the University of Virginia, was my graduate student at Stanford, and together we began to ask that question. As a first step we needed a Siren-like tempter appropriate for preschoolers in the Surprise Room. He or she or it had to meet two criteria: the tempter had to be seductive but also considered acceptable to parents, the Bing Nursery School director, and the researchers, in addition to my three young daughters, who served as my advisory board. The result was Mr. Clown Box (below).

Mr. Clown Box was a large wooden box with a brightly painted clown's face. The smiling face was surrounded by blinking lights and flanked by outstretched arms, each appearing to hold up a glass-windowed compartment. When the compartment lights turned on, tempting little toys and treats rotated very slowly on a drum inside each window. Mr. Clown Box was a big talker and a powerful tempter. A speaker hidden in his head was connected to a tape recorder and microphone in the observation room.

We wanted to simulate a situation that everyone faces repeatedly in life: when you have to resist powerful immediate temptations and seductions for the sake of more important but delayed outcomes. Think of the teenager trying to complete overdue homework who is asked to join his best friends for a movie, or the happily married older executive invited for drinks by the attractive young personal assistant after a long day together at the annual sales convention, far from home. Mr. Clown Box served as the seducer who would greatly tempt young kids.

During these studies, Charlotte played briefly with the child — in this example, "Sol," age four — in a corner of the Surprise Room that contained both attractive and broken toys. She then seated Sol at a small table facing Mr. Clown Box. She explained that she would have to leave the room for a while, and she showed Sol his "job." He had to work the whole time, without interruption, on a particularly boring task. For example, he might have to copy the squares from a worksheet filled with either *X*'s or *O*'s into the adjacent empty squares on the same worksheet, or put little pegs from a large pile into a Peg-Board. If he did that without interruption, then he could play with the fun toys *and* Mr. Clown Box when Charlotte returned;

otherwise, he would be able to play with only the broken toys. She emphasized that he had to work the whole time she was out of the room in order to finish his job, and he solemnly promised he would do so. She carefully forewarned Sol that Mr. Clown Box might try hard to play with him, and she stressed that looking at, talking to, or playing with him would make it impossible for Sol to finish his job.

Charlotte then invited Sol to meet Mr. Clown Box, who lit up brightly, flashing his lights and illuminating his toy-filled windows, and introduced himself in a loud and pleasant voice: "Hi! I'm Mr. Clown Box. I have big ears and I love it when children fill them with all the things they think and feel, no matter what." (He had obviously had at least some training in psychotherapy.) Mr. Clown Box "ahem"ed and "aha"ed encouragingly in response to whatever Sol said and engaged him in a brief, pleasant conversation in which he invited Sol to play with him. He demonstrated that a distinctive *bzzt* sound indicated that he was about to do something fun that Sol would want to watch, and briefly lit his display windows, letting Sol glimpse the attractive toys and treats rotating slowly within them.

A minute after Charlotte exited, Mr. Clown Box lit up, flashed his lights, and laughed: "Ho, ho, ho, ho! I love to have children play with me. Will you play with me? Just come over and push my nose and see what happens. Oh please, won't you push my nose?"

For the next ten minutes he continued his tortures, mercilessly tempting the child, his lights turning on and off around his face and in his display windows, a bright light on his bowtie also flashing. He resumed his seductive efforts every 1½ minutes:

"Oh, I'm having such a good time! I'll make even more fun for us if you just put down your pencil. Put down your pencil and then we'll really have a good time. Please put down your pencil and come play with me.... Just come over and push my nose, and then I'll do tricks for you. Wouldn't you like to see some of my surprises? Look in my windows now."

Eleven minutes after Charlotte's exit, Mr. Clown Box turned off and she returned to the Surprise Room.

IF-THEN PLANS FOR RESISTING TEMPTATION

For preschoolers, the clown was probably as tough to resist as the Sirens were for Odysseus, and unlike the Greek hero tied to his mast, the kids were not bound to their chairs, nor did they have beeswax in their ears like his crew. Our question was: what could help preschoolers like Sol better resist the temptations that Mr. Clown Box would use to lure them?

Guided by the marshmallow findings, we figured that to effectively resist a hot temptation (whether to eat the marshmallow now or cave in to any other temptation), the inhibitory *No!* response had to replace the hot *Go!* response — and it had to do this quickly and automatically, like a reflex. All you needed, in the language of Hollywood's movie industry, was a good connection, one that created an automatic link between the needed *No!* response and the hot stimulus (which normally triggered *Go!*). For example, one temptation-inhibiting plan might be to instruct the preschooler as follows:

"Let's try to think of some things that you could do to keep yourself working and not let Mr. Clown Box slow you down.

Let's see . . . One thing you can do is this: when Mr. Clown Box makes that *bzzt* sound and asks you to look at him and play with him, you can just look at your work, not him, and say, 'No, I can't; I'm working.' And when you say it, do it. He says, 'Look,' and you say, 'No, I can't; I'm working.' "

This type of *If-Then* plan specifies the tempting hot stimulus — "*When* Mr. Clown Box says to look at him and play with him" — and links it to the desired temptation-resisting response: "*then* you can just *not* look at him and say, 'I'm *not* going to look at Mr. Clown Box.' " Preschoolers armed with this type of plan reduced their distraction time and kept on working, with the best results. Even when the clown succeeded in distracting them from their work, the disruption lasted on average less than five seconds, and the children inserted an average of 138 pegs into their Peg-Boards. In contrast, those without this type of plan interrupted their work for an average of 24 seconds per distraction and inserted only 97 pegs. In the preschool world of peg insertion, these were big differences. We also saw that many children who received this plan instruction innovated their own variations ("Quit that!" "Stop that!" "Dummy!"), which let them get their pegs into the holes faster, and ultimately allowed them to play happily with Mr. Clown Box as well as with the unbroken toys.

Our research with Mr. Clown Box turned out to be the opening step for an important independent research program developed many years later by Peter Gollwitzer, Gabriele Oettingen, and their colleagues at New York University. Beginning in the 1990s, they identified simple but surprisingly powerful *If-Then* plans for helping people deal more effectively with a wide variety of otherwise crippling self-control problems — even under very difficult and emotionally hot conditions, when they were trying to pursue important but hard to achieve goals. Now called *If-Then* implementation plans, these plans have helped students study in the midst of intrusive temptations and distractions, aided dieters in forgoing their favorite snacks, and enabled children with attention deficit disorders to inhibit inappropriate impulsive responses.

TAKING THE EFFORT OUT OF EFFORTFUL CONTROL

With practice, the desired action of an implementation plan becomes initiated automatically when the relevant situational cues occur: When the clock hits 5 p.m., I will read my textbook; I will start writing the paper the day after Christmas; when the dessert menu is served, I will not order the chocolate cake; whenever the distraction arises, I will ignore it. And implementation plans work not just when the *If* is in the external environment (when the alarm rings, when I enter the bar) but also when the cue is your internal state (when I'm craving something, when I'm bored, when I'm anxious, when I'm angry). It sounds simple, and it is. By forming and practicing implementation plans, you can make your hot system reflexively trigger the desired response whenever the cue occurs. Over time, a new association or habit is formed, like brushing teeth before going to bed.

Such *If-Then* plans, when they become automatic, take the effort out of effortful control: you can trick the hot system into reflexively and unconsciously doing the work for you. The hot system then lets you automatically act out the script you want when you need it, while your cool system rests. But unless you incorporate the resistance plan into the hot system, it is unlikely to be activated when you need it most. That is because emotional arousal and stress increase when you are faced with hot temptations, thus accelerating the hot system, triggering the quick, automatic *Go!* response, and attenuating the cool system. When the hot temptation arrives — whether from Mr. Clown Box flashing his lights in the Surprise Room, the chocolate dessert on the menu, or the attractive colleague in the bar at the business convention — the automatic *Go!* response is likely to win if there is no well-established *If-Then* plan. However, when *If-Then* plans are established, they work well in surprisingly diverse settings, with different populations and age groups, and they can help people more effectively achieve difficult goals — goals that they previously thought they could not reach.

One impressive example is with children with attention deficit/hyperactivity disorder (ADHD). ADHD is an increasingly common problem, and children who have it often experience many academic and interpersonal troubles. They are highly vulnerable to distraction and tend to have difficulty controlling their attention, making it hard to stay task-oriented. These cognitive limitations can undermine children with ADHD in many academic and social situations, causing stigmatization and the risk of overmedication. *If-Then* implementation plans have helped such children solve math problems faster, substantially improve on tasks assessing working memory, and persevere in their efforts to resist distraction under very difficult laboratory conditions. These applications illustrate the power and value of implementation plans, and paint an optimistic picture of the human potential for self-generated change. The continuing challenge is to translate these procedures from short-term experiments into long-term intervention programs that produce sustained change in everyday life.

6

IDLE GRASSHOPPERS AND BUSY ANTS

THE MARSHMALLOW EXPERIMENTS ALLOWED us to see how children managed to delay and resist temptation, and how differences in this ability play out over a lifetime. But what about the choice itself?

I started to ask that question while I was a graduate student at Ohio State University, well before I joined the Stanford faculty. I spent one summer living near a small village in the southern tip of Trinidad. The inhabitants in this part of the island were of either African or East Indian descent, their ancestors having arrived as either slaves or indentured servants. Each group lived peacefully in its own enclave, on different sides of the same long dirt road that divided their homes.

As I got to know my neighbors I became fascinated by what they told me about their lives. I discovered a recurrent theme in how they characterized each other. According to the East Indians, the Africans were just pleasure-bent, impulsive, and eager to have a good time and live in the moment, while never planning or thinking ahead about the future. The Africans saw their East Indian neighbors as always working and slaving for the future, stuffing their money under the mattress without ever enjoying life. Their descriptions were strikingly reminiscent of Aesop's classic fable about the grasshopper and the ant. The indolent, hedonistic grasshopper is hopping around, chirping happily in the summer sun and enjoying the here and now, while the worried, busy ant is toiling to put away food for the winter. The grasshopper is indulging in hot-system pleasures, while the ant is delaying gratification for the sake of later survival.

Did the road that separated the two groups in this village really divide self-indulgent grasshopper types from future-oriented, hardworking ants? To check if the perceptions about the differences between the ethnic groups were accurate, I walked down the long dirt road to the local school, which was attended by children from both groups. The school was still run by the British colonial educational system, and the children were dressed in fresh white shirts or blouses. Everything seemed neat, proper, and orderly, and the children waited with clasped hands for the teacher to begin the lesson.

The teachers welcomed me into their classrooms, where I tested boys and girls between the ages of 11 and 14. I asked the children who lived in their home, gauged their trust that promises made would be promises kept, and assessed their achievement motivation, social responsibility, and intelligence. At the end of each of these sessions, I gave them choices between little treats: either one tiny chocolate that they could have immediately or a much bigger one that they could get the following week. During the session they also chose between receiving $10 right then or $30 if they waited a month, and between a "much larger gift much later or a smaller one now."

The young adolescents in Trinidad who most frequently chose the immediate smaller rewards, in contrast to those who chose the delayed larger ones, were more often in trouble and, in the language of the time, judged to be "juvenile delinquents." Consistently, they were seen as less socially responsible, and they had often already had serious issues with authorities and the police. They also scored much lower on a standard test of achievement motivation and showed less ambition in the goals they had for themselves for the future.

TRUST

Consistent with the stereotypes I heard from their parents, the African Trinidadian kids generally preferred the immediate rewards, and those from East Indian families chose the delayed ones much more often. But surely there was more to the story. Perhaps those who came from homes with absent fathers — a common occurrence at that time in the African families in Trinidad, while very rare for the East Indians — had fewer experiences with men who kept their promises. If so, they would have less trust that the stranger — me — would ever really show up later with the promised delayed reward. There's no good reason for anyone to forgo the "now" unless there is trust that the "later" will materialize. In fact, when I compared the two ethnic groups by looking only at children who had a man living in the household, the differences between the groups disappeared.

Beginning in early childhood, far too many people live in untrustworthy, unreliable worlds in which promises for delayed larger rewards are made but never kept. Given this history, it makes little sense to wait rather than grab whatever is at hand. When preschoolers have an experience with a promise maker who fails to keep his promise, not surprisingly they are much less likely to be willing to wait for two marshmallows than to take one now. These commonsense expectations have long been confirmed in experiments demonstrating that when people don't expect delayed rewards to be delivered, they behave rationally and won't choose to wait for them.

A few years after my time in Trinidad, before the marshmallow experiments, I was teaching at Harvard and I continued to study such choices among children and adolescents in Cambridge and Boston. The year 1960 was an odd moment in history to be studying delay of gratification and self-control in Harvard University's Social Relations Department. Much was changing. Timothy Leary had joined the faculty and was experimenting with the "magic mushrooms" he had found on a trip to Mexico, attempting to create new psychedelic, mind-altering experiences not just in himself but also in our students. One morning, mattresses suddenly replaced several of the graduate student desks, and large packages from a chemical company in Switzerland began to arrive in the department mail. With the help of the drug LSD, the age of "Turn on, tune in, drop out" had begun. Leary was leading the charge into the counterculture, and many of our graduate students followed him.

As much of the world seemed to be losing self-control, it felt particularly timely to keep studying it. Carol Gilligan, who was working toward a doctoral degree, and I collaborated on a new project, testing sixth-grade boys from two public schools in the Boston area. We wanted to see if children who consistently chose to wait for larger but delayed rewards, rather than immediately available, albeit smaller, ones, would be more able than Adam and Eve to resist a strong temptation when they ran into it. But 12-year-old schoolboys in Boston needed something more tempting than an apple.

In a first session in their classrooms, the boys completed a variety of tasks, and we offered them many choices between smaller-now versus bigger-later rewards to thank them, much as I had in Trinidad. We wanted to see if their preferences for delayed-larger versus immediate-smaller rewards would be linked to how they dealt with powerful temptations in a new situation. Would those who delayed more in the first session be less likely to give in to a strong temptation in a different situation — one in which cheating was the only way to succeed?

To answer this question, we set up seemingly unconnected individual sessions later in the semester, during which we introduced each child, one at a time, to a game of skill. Ostensibly, the purpose of the game was to see how effectively and quickly each boy could use a "ray gun" to destroy a rocket that had become disabled in the space race against the Soviet Union (this was big news at the time). The large toy ray gun was painted silver, mounted on a plank, and pointed toward a high-speed "rocket" target. Above the target, a row of five lights illuminated the number of points earned after each shot. Three brightly colored sportsmen badges (marksman, sharpshooter, and expert) were flashed and offered as prizes, to be awarded on the basis of the total number of points obtained. Although any young kid today would dismiss this 1960s ray gun as a quaint museum piece, at that time the 12-year-old boys found it irresistible.

"Let's pretend that the rocket is disabled and must be

destroyed," Carol would say. "To those of you who are good shots, I'm going to give this marksman badge; to those of you who are better than marksmen, I'm going to give this sharpshooter badge; and to those of you who are really good — better than marksmen or sharpshooters — I'm going to give this expert badge."

Unbeknownst to the boys, the number of points scored for each shot was random and bore no relation to their skill level, and the scores they got made it impossible to win a badge: to get a badge, they had to cheat by falsifying their scores, and to win a better badge they had to fake it even more. The boys kept their own scores while playing alone in the room, and both the timing and amount of their cheating was computed. The results were clear: the same patterns we saw in Trinidad with the "juvenile delinquents" choosing smaller immediate rewards held in Boston. Those who had consistently chosen to wait for larger but delayed rewards rather than immediate but smaller rewards in the earlier session cheated much less than those who had taken the smaller rewards. If the boys who had preferred the delayed rewards did cheat, they waited much longer before they gave in to the temptation to falsely inflate the scores they reported.

HOT NOW VERSUS COOL LATER: THE VIEW FROM THE BRAIN

In 2004, half a century after my neighbors in Trinidad had described each other as if they were either Aesop's happy grasshoppers or hardworking ants, I was excited to read a study by Samuel McClure and his colleagues in the journal *Science.*

These researchers had taken a step forward in analyzing how people make decisions: they used fMRI machines to study what went on in the brain when people chose between getting rewards in the here and now versus in the future.

Psychologists and economists often note that people tend to be highly impatient and driven mostly by the hot system when they deal with immediate rewards — but they can be patient, rational, and cool in their preferences when they choose between rewards that are all delayed. While this inconsistency has long been recognized, the brain mechanisms that underlie it remained a puzzle. To try to solve it, McClure and his team began with the hypothesis about the role of the hot and cool systems in the brain. They reasoned that the emotional hot (limbic) system underlies short-run impatience: it is activated automatically by immediate rewards and triggers the *Go!* response of "I want it now!" It is relatively insensitive to the value of delayed rewards or anything in the future. In contrast, long-lasting patience of the kind needed to choose rationally between different delayed rewards, for example in retirement planning, depends on the cool cognitive system — particularly on specific areas in the prefrontal cortex and other closely related structures that developed much later in the course of human evolution.

McClure's team gave adults choices between monetary options whose expected delivery times varied from "now" to a later time (for example, $10 now or $11 tomorrow), as well as choices between rewards that were all in the future, like $10 in a year or $11 in a year and a day. The researchers used fMRI to monitor the neural regions of the hot and cool systems of each participant. As participants made their decisions, the researchers discovered that the degree to which each neural region became engaged would predict whether the individual chose a smaller immediate payoff or a larger delayed payoff: neural activity occurred in the hot region when participants were choosing between the two near-term rewards (an amount today versus slightly more tomorrow) and in the cool region when they were choosing between future rewards (an amount in a year versus slightly more in a year and a day). McClure and his colleagues thus confirmed that there are in fact two neural systems — one hot and the other cool — that evaluate immediate versus delayed rewards separately. For me, it was reassuring to see that the activity in the brain turned out to be consistent with what we had inferred from the preschoolers' behavior in the Surprise Room. In 2010, another group of researchers, headed by Elke Weber and Bernd Figner at Columbia University, conducted an experiment that located more precisely the specific brain region that lets us choose to wait for delayed rewards: it is the left, but not the right, lateral prefrontal cortex.

Immediate rewards activate the hot, automatic, reflexive, unconscious limbic system, which pays little attention to delayed consequences. It wants what it wants immediately and steeply reduces or "discounts" the value of any rewards that are delayed. It is driven by the sight, sound, smell, taste, and touch of the object of desire, whether it's marshmallows that make preschoolers ring the bell, irresistible fudge cake on the dessert platter, or the Siren songs that drowned sailors in an ancient myth. That's why smart people in the public eye, like presidents, senators, governors, and financial tycoons, can make stupid decisions when immediate temptations lure them to overlook the delayed consequences.

Delayed rewards, in contrast, activate the cool system: the

slower-to-respond but thoughtful, rational, problem-solving areas of the brain's prefrontal cortex that make us distinctively human and able to consider long-term consequences. As we've seen in earlier chapters, delay ability can help us slow down and "cool it" long enough for the cool system to monitor and regulate what the hot system is doing. To reiterate, the two systems — one hot to deal with immediate rewards and threats, the other cool to deal with delayed consequences — act together: as one becomes more active, the other becomes less active. The challenge is to know when it's best to let the hot system guide your course, and when (and how) to get the cool system to wake up.

McClure and his colleagues also invoked Aesop's classic fable to summarize their conclusion: "Human behavior is often governed by a competition between lower level, automatic processes that may reflect evolutionary adaptations to particular environments, and the more recently evolved, uniquely human capacity for abstract, domain-general reasoning and future planning. . . . The idiosyncrasies of human preferences seem to reflect a competition between the impetuous limbic grasshopper and the provident prefrontal ant within each of us."

We may all be both grasshopper and ant, but whether the prefrontal ant or the limbic grasshopper in us emerges at any given time depends on the temptation in the particular situation and how we appraise and think about it. As Oscar Wilde famously noted, "I can resist everything except temptation."

7

IS IT PREWIRED? THE NEW GENETICS

Born in Chicago in 1928, James grew up worrying about his maternal Irish heritage. His goal was to be the smartest kid in the class, at a time when the Irish in Chicago were often the butt of intelligence jokes. He remembers hearing stories as a child about ads for job openings that ended with "No Irish Need Apply." James recognized that although he clearly had strong Irish genes, there was no evidence that he was on the slow side. Fortunately, he concluded that "the Irish intellect, and the shortcomings for which it was known, must have been shaped by the Irish environment, not by those genes: nurture, not nature, was to blame." James, whose last name is Watson, and Francis Crick were awarded the Nobel Prize in 1962 for discovering the structure of DNA. It opened a window into a new understanding of who we are and what we can be. In the half century since James D. Watson shook hands with the King of Sweden, the astonishing answers have kept coming.

In 1955, around the same time that Watson and Crick were

working on the structure of DNA, "Mr. Abe Brown" brought his ten-year-old son, "Joe," to the psychology department's clinic at Ohio State University, where I was a trainee in the doctoral program. Mr. Brown seemed in a big hurry and wasted no time on preliminaries before blurting out his one question about Joe, who was seated next to him: "I just want to know: is he stupid or is he just lazy?"

Mr. Brown's blunt question reflects the same concern raised (typically more diplomatically) by anxious parents after every talk I give about the Marshmallow Test. It's the same question the young James Watson tortured himself with and was smart enough to answer for himself. It's a question, often *the* question, when my talks turn to the causes of human behavior: *Is it nature, or is it nurture?* In the first few minutes I spent with Mr. Brown, his implicit theory about nature and nurture became clear. If Joe was stupid, there was nothing Mr. Brown felt he could do about it, and he would try to accept it and ease up on his son. On the other hand, if Joe was "just lazy," Mr. Brown had strong ideas about the kind of discipline he would use to help "shape him up."

For centuries, the argument about the genetic versus environmental influences on the brain and behavior has raged with virtually every important human characteristic, from the origins of intelligence, aptitudes, and abilities; to aggression, altruism, conscientiousness, criminality, willpower, and political beliefs; to schizophrenia, depression, and longevity. The arguing has not been confined to the academic battleground. It influences thinking about social policy, politics, economics, education, and child rearing. How we vote on policy issues, for example, is influenced by whether we attribute economic and achievement inequalities primarily to genetic or environmental forces. If the differences are due to nature, society may decide to take pity on the unfortunates who lost out in the genetic roulette that produced them but could also feel that the rest of the world is not culpable for their misfortune. If it's the environment that is mostly responsible for who we are and what we become, then is it up to us to change it to reduce the injustices it has produced? How you see the role of heredity and prewiring in willpower, character, and personality affects not just your abstract view of human nature and responsibility, but also your sense of what is and is not possible for you or your children.

The accepted scientific views on nature-nurture have reached diametrically opposed conclusions at different points in my lifetime. In the behaviorism that dominated American psychology into the 1950s, scientists like B. F. Skinner saw newborns as entering the world as if they were blank slates, ready to have the environment stamp itself on them to determine what they became and shape them mostly through the rewards or reinforcement it delivered. Beginning in the 1960s, such extreme environmentalism receded. By the 1970s, thinking on this topic was transformed, as Noam Chomsky and many other linguists and cognitive scientists proved that much of what makes us human is prewired. The initial battle was fought over how babies acquire language. The winners showed that the underlying grammar that enables language is largely innate, although whether the baby ultimately speaks High German or Mandarin of course depends on learning and the social environment. The newborn's slate, far from being blank is deeply encrypted.

The list of what babies bring with them from the womb grows longer and more amazing each year. Elizabeth Spelke at

Harvard University is one of the leaders looking into the baby's mind and brain, using the infant's gaze as a tool to see what he does and doesn't understand. She tells us, for example, that babies are born accountants, with a remarkable readiness to understand numbers and all set to do geometry — at least when it comes to figuring out how to navigate in three-dimensional space to discover hidden treasures. What infants are equipped to understand seems limited mostly by our ability as adults to figure it out.

TEMPERAMENTS

Parents have long recognized that their babies differ greatly in temperament, and they see these innate differences in emotional reactions soon after birth. These differences were captured in the ancient Greco-Roman typology that linked innate emotional dispositions to four vital body humors, which served as the early version of DNA. In that theory, when blood was predominant, the person was *sanguine*, characterized as good-natured and cheerful; black bile underlay the *melancholic* individual who tended to be anxious and moody; a readiness to be angry and irritable, due to too much yellow bile, marked the *choleric*; and when phlegm predominated, the person was *phlegmatic*, or easygoing and slow to become aroused.

Babies enter the world with physiological differences in their emotional reactivity, activity level, and ability to control and regulate their attention. Although these differences begin with prewiring, by the time babies are born, they have already been sculpted for many months by their uterine environment. These differences significantly influence what they feel, think, and do and who they become — including how easily they exert self-control and delay gratification. New parents eagerly, and often wearily, discuss how their baby's temperament has transformed their lives. It's not news that most babies are a mix of everything emotionally. At the extremes, some babies are very active, smile and laugh a lot, and show intense pleasure early in life. Others are highly emotional, readily aroused, and prone to negative affect; these babies are often distressed, irritable, and angry, especially when frustrated (which seems to be much of the time). Babies also vary in sociability. Some are fearful when exposed to strangers or even new toys, while others seem to be eager to interact with anything and anybody. Some rarely feel fear, but are terribly afraid and hard to console when they do; others are often mildly fearful but rarely terrified.

Babies vary in the vigor or intensity of their responses and their tempo and speed, ranging from those who sleep a lot (and let others do the same) to those who are busy with action and connection seeking, regardless of the time of day or night. These temperamental differences are visible not just in how active, easy, happy, distressed, or glad to be alive the baby seems to be, but also in how much smiling, laughing, playing, and sleeping the parents are doing, and how much joy rather than exhaustion and desperation they feel. The emotional behavior of children continuously influences that of their caregivers, and vice versa, escalating into more pleasure on one end of the continuum and more distress on the other.

Emotional dispositions also influence how well, how soon, and under what conditions different children do or don't manage to regulate their attention, delay gratification, and exert self-control as they develop over the years. How heritable are these emotional characteristics? Most people who ask the question realize, after a moment's reflection, that surely the answer must be a combination of both heredity and environment. For many years, studies with twins — especially pairs of identical twins, who begin life as genetically similar as two people can be — compared those who were reared together within the same family with those who grew up apart in different families. These comparisons were used to try to determine the different effects of nature and nurture on behavioral dispositions and psychological characteristics. The details are continuously disputed, but a reasonable estimate from twin research is that about one-third to one-half of whatever develops can be attributed to genetic variation. In the case of intelligence, some of the similarity estimates for identical twins have been even higher. It is notable, however, that even with identical twins reared together, it is entirely possible for one to develop schizophrenia, severe depression, or other mental or physical illnesses, while the other lives out a healthy life. It is also possible for one to become a model of high self-control, while the other personifies impulsiveness.

Researchers have used twin studies to parse nature and nurture into distinct percentage contributions, as if they were separable. We should be grateful for their pioneering work, which made it clear at last that we are biological creatures, heavily prewired, and that nature matters as much as nurture. But as the research on heritability deepens, we see that nature and nurture are not easily separated. Human dispositions and behavior patterns, including character and personality, attitudes, and political beliefs, reflect the complex effects of genes (usually multiple genes) whose expressions are shaped by environmental determinants throughout the course of life. Who we are and what we become reflects the interplay of both genetic and environmental influences in an enormously complex choreography. It is time to put away the "How much?" question because it cannot be answered simply. As the Canadian psychologist Donald Hebb noted long ago, it's like asking, What's the more important determinant of a rectangle's size: its length or its width?

UNPACKING THE DNA LIBRARY

The conclusion is inescapable: who we are emerges from a tightly intertwined dance between our environment and our genes that simply can't be reduced to either part alone. But the unraveling mysteries of DNA, from breaking the code to sequencing the entire human genome and mapping its many regulatory elements, have begun to provide a molecular basis through which our "nurture" interacts with our "nature" to make us who we are.

DNA is a biological code that provides instructions that enable our cells to make and do all that is needed for life. In the human body, each of approximately a trillion cells holds within its nucleus a complete and identical sequence of DNA. That is about 1.5 gigabytes of genetic information, and it would fill two CD-ROMs, yet the DNA sequence itself would fit on the point of a well-sharpened pencil.

If this sounds like a lot, it is only the tip of the iceberg, because the real flexibility and complexity are in how the DNA is organized and used. The DNA "letters" of the code — A, C, G, and T — can be combined in unique and varied ways into different "words." More important, increasingly complex levels of organization, including how, where, and when the "words" are put together, allow for the vast repertoire of individual differences that make us unique. How does this work?

Consider all the information contained within a library that houses thousands of books as a metaphor for the human body, which "houses" about twenty thousand genes. Each book in this DNA library contains words arranged into sentences. These DNA sentences are genes. The sentences are further organized into paragraphs and chapters. These are modules of highly coordinated genes that function together, which are further organized into books, which are further organized into sections of the library (tissues, organs, etc.). Here is the critical piece: the overall "experience" of the book reader visiting the library is not simply the sum of all the books in the library. The experience of the reader depends on when he visits the library, who joins him, what sections he visits, what parts of the library are open or closed at that particular time, and which books he pulls off the shelves. In short, what gets read, the genes that will and won't be expressed, depends on the enormously complex interactions between biological and environmental influences. The possibilities are endless and the role of the environment essential. Our genetic makeup (i.e., our library) provides a stunningly nimble system for responding to the environment.

The puzzle is figuring out the physical properties of DNA that allow for this responsiveness to the environment. It turns out that a relatively small fraction of DNA encodes words arranged into sentences (i.e., the genes). The bulk of DNA residing between these sentences was long considered to be noncoding "junk" whose function was a mystery. Recent work is beginning to show that these long stretches of noncoding DNA are not junk at all. Instead, they are of central importance in determining how our DNA is expressed. The junk is filled with critical regulatory switches that decide what sentences are made — and when, where, and how they are made — in response to cues that come from the environment. Given these discoveries, Frances Champagne, a leader in research on how environments influence gene expression, is convinced that it is time to drop the nature versus nurture debate about which is more important and ask instead, What do genes actually do? What is the environment doing that changes what the genes do?

Ultimately, all biological processes are influenced by context, including the social-psychological environment. The environment includes everything from the mother's breast milk, the broccoli or bacon eaten, the drugs consumed, and the toxins absorbed, to the social interactions, stresses, defeats, triumphs, elations, and depressions experienced over the course of a life. And the environment is most influential early in life. For example, already during gestation, the stress experienced by mothers exposed to violence by their intimate partners may be transmitted to their offspring, making the babies more vulnerable to serious behavioral problems even much later in life. Stress in childhood influences gene expression in many but by no means all children, and tends to induce a defensive reaction, characterized by heightened immune and stress reactivity. These results suggest that the cellular milieu of the baby's brain is profoundly influenced by the maternal environment.

And remarkably, the environmental influences may even precede conception. Flying in the face of previous beliefs about heritability, recent evidence indicates that some non-genomic characteristics of our cells are inherited. At a molecular level, notes Champagne, this means that these characteristics, induced by the social and physical environment, can alter the characteristics of the cells that ultimately create the individual's offspring. The details of how this happens are just beginning to be uncovered. But the sobering message is that the inheritance of both risk and resilience in dealing with social interactions may be passed on across generations. This implies that the way young adolescents and adults live their lives, what they eat, drink, and smoke, and the joys and stresses of their social interactions and experiences may in part shape what will be expressed or left unread in the genomes of their offspring.

In the first year of life, the prefrontal cortex begins to develop in ways that are essential for self-control and self-generated change. In the metaphor of the hot and cool systems, this marks the beginning of the cool system that in time slowly enables self-control. Between the ages of about three and seven, this development increasingly allows children to shift and focus their attention, to regulate emotions adaptively, and to inhibit unwanted responses in order to pursue their goals more effectively.

These changes allow a child to begin to regulate her own feelings and reactions as she gets older, and to modify how her prewiring plays out, rather than being its victim. This ability to self-regulate in ways that can change how predispositions are expressed is captured in an anecdote told by Harvard's Jerome Kagan, the leading figure in shyness research. When his granddaughter was in preschool and struggling to overcome her shyness, she asked him to pretend he did not know her so she could practice not being shy — and in time, it worked for her. Kagan's earlier research had made it clear that while a predisposition like shyness has genetic roots, it is also amenable to change. Good preschool experiences and caretakers who manage to overcome their overprotectiveness can help the shy child become less timid. Kagan's granddaughter showed the preeminent shyness researcher that the child herself can be an active agent in her own development and use diverse strategies to change how her dispositions play out in her life.

WHAT THE RODENTS DO

Rodents, long shunned and exterminated in the home, are popular subjects for experiments on the role of nature and nurture because the mouse genome is amazingly close to our own. Seeing what mice and other rodents do can answer questions about human behavior that cannot be tested with people. In 2003, a team of researchers at Emory University, led by Thomas Insel and Darlene Francis, used two strains of mice (BALB and B6) that differ greatly in their novelty seeking and fearfulness. BALB mice are bred to be genetically shy, making them fearful in their behavior and causing them to hide in a corner of their cage. They are sharply different from B6 mice, who are genetically bred to seek novelty and become relatively fearless. The researchers tested how the genetically brave, novelty-seeking mice would behave when placed in an environment with a shy, fearful mother. The genetically brave mice placed with shy mothers became more similar to genetically shy mice who had grown up with their own shy mothers. Two clear conclusions emerge. First, genetic endowment is an important determinant of behavior. Just as important, however, is the maternal environment early in life. It has a powerful impact on how those genes function.

In one study published in the 1958 *Canadian Journal of*

Psychology, researchers used rats that had been selectively bred to be either "maze-dull" or "maze-bright." Over multiple generations, this selective breeding produced rats that were primed to be bright or dull when it came to running mazes. The scientists placed these young animals either in a very lively rat world filled with many sensory stimuli or in an impoverished, barren rat world that had essentially no sensory stimuli. The dull rats put into the enriched environments became significantly brighter, and the bright ones stuck in the impoverished life space got duller, showing a significant decline in their performance. The environment dramatically changed the expression of a cognitive ability that was generated through selective breeding to create rats as genetically bright or dull as their genes could possibly make them. This study was one of the first to demonstrate that what genes do depends on the environments in which they are functioning.

Mothers and other caregivers vary enormously in how much they nurture their young, but there is no way to manipulate or control these effects in experiments with people. Consequently, rodents were used again in another study to see if the stimulation that rat mothers gave their pups early in life changed what their offspring became. When rat mothers have pups, they lick and groom them, but there are large, stable differences in how much licking and grooming (abbreviated "LG") they give their pups. Some lick and groom at a much greater rate than others, just as some human mothers give much more stimulation and affection to their babies than others. The study showed that the rat pups fortunate enough to have high LG mothers benefited greatly. They performed better on cognitive tasks and had lower physiological arousal responses to acute stress than those who were stuck with low LG mothers.

James R. Flynn, a New Zealand psychologist, discovered a general upward trend in IQ scores not in rodents but in our own species, across industrialized countries like the United States and Britain. There were significant increases in scores from one generation to the next. On measures of intelligence that required problem solving and did not rely on verbal knowledge and symbols, average increases of about 15 points per generation were found. One thing is certain: in the sixty years these studies cover, the changes are surely not due to evolution and cannot be attributed to genetic changes in the population. This is encouraging evidence of the power of the environment to influence characteristics like intelligence. Even if traits like intelligence have large genetic determinants, they are still substantially malleable. James Watson summarizes the conclusion: "A predisposition does not a predetermination make."

Compelling examples of human gene and environment interplay come from a study in New Zealand of more than one thousand children followed from birth in 1972 for more than 30 years. Researchers tested to see if the number of stressful life events experienced over a 20-year period influenced the long-term risk of depression. Concurrently, they assessed participants for variation in a gene that alters the level of serotonin in the brain. Again, it was the interplay of genetics and environment that mattered and determined whether or not the genetic potential for risk or resilience was activated. Depression emerged more often in people who had the genetic vulnerability *if* they were also exposed to more stressful life experiences.

GETTING OVER "NATURE VERSUS NURTURE"

Our genes influence how we deal with the environment. The environment affects which parts of our DNA are expressed and which are ignored. What we do, and how well we control our attention in the service of our goals, becomes part of the environment that we help create and that in turn influences us. This mutual influence shapes who and what we become, from our physical and mental health to the quality and length of our life.

To reiterate, human dispositions and behavior patterns, including character and personality, attitudes, and even political beliefs, reflect the complex effects of our genes, whose expressions throughout the course of life are shaped by a host of environmental determinants. Dispositions are produced by the interplay of both genetic and environmental influences in an enormously complex choreography — and that means it's time to get over the nurture *or* nature question. As Daniela Kaufer and Darlene Francis (at the University of California, Berkeley) concluded in 2011, the findings from cutting-edge research on the nature-nurture relationship "are inverting implicit assumptions about gene-environment relationships.... Environments can be as deterministic as we once believed only genes could be and... the genome can be as malleable as we once believed only environments could be."

To answer Mr. Brown's question about Joe half a century after he asked it, most predispositions are prewired to some degree, but they are also flexible, with plasticity and potential for change. Identifying the conditions and mechanisms that enable the change is the challenge. I think Mr. Brown would not be pleased with this response. His hot system wanted a quick one-word answer: dumb or lazy. But the more we learn about nature and nurture, the more it is clear that they inseparably shape each other.

PART II

FROM MARSHMALLOWS IN PRE-K TO MONEY IN 401(k)

In Part I we saw how preschoolers manage to delay gratification, and how the skills that enable self-control can be enhanced and nurtured. While much of what makes self-control less effortful is prewired, much of it remains open to learning. The cognitive and emotional skills that make it possible for preschoolers to wait for a bigger reward pave the way for them to develop the psychological resources, mind-sets, and social relationships that can improve their chances to build the fulfilling and successful lives they want. In Part II, I look at how this works and how the ability to delay gratification protects the self by helping people control and regulate their personal vulnerabilities more effectively, cool their hot impulsive responses, and take consequences into account. I examine the journey from preschool to the rest of life, parsing the underlying connections between the seconds that children spent waiting for more marshmallows in pre-K and how they were faring at midlife. If we understand these connections, we can develop them and learn how to better help our children and ourselves.

To begin with, the hot system deserves to be appreciated, listened to, and learned from. It gives us the emotions and zest that make life worth living and allows automatic judgments and decisions that work well some of the time. But the hot system has its costs: it effortlessly makes quick judgments that feel right intuitively but are often dead wrong. It can save your life by getting you to hit the car brakes in time to avoid a collision or to duck for cover when you hear a gunshot nearby, but it can also get you into trouble. It can cause well-intentioned policemen to shoot too quickly at innocent but suspicious-looking strangers in dark alleys, drive loving couples apart with jealousy and mistrust, or lead overconfident high achievers to wreck their lives with impulsive greed or fear-driven decisions. And its excesses — the temptations it dangles that one can't resist, the fears it too vividly creates, the stereotypes it triggers from minimal information, and the conclusions and decisions it pushes us to make too quickly — can be hazardous to health, wealth, and well-being. Part II explores some of these risks and possible ways to control them and perhaps even learn from them.

Natural selection shaped the hot system to enable survival and the spread of human DNA in a tough Darwinian world, but much later in the course of evolution, it also created the cool system. The cool system gives humans the ability to behave intelligently, with imagination, empathy, foresight, and sometimes even wisdom. It allows us to reappraise and reconstrue the meaning of events, situations, people, and our lives. The ability to think in constructive, alternative ways can change the impact of stimuli and life events on what we feel, think, and do, as the preschoolers in Part I demonstrated. Therein lies the potential for being purposeful agents of our actions, for taking charge, for exerting control, and for influencing how life plays out.

The mental mechanisms that enable self-control in the face of temptation also play a crucial role in efforts to regulate and cool down painful emotions, like heartbreak and interpersonal rejection. These mechanisms are supported by the psychological immune system, which works ingeniously to protect self-regard, reduce stress, and make most of us feel good — or at least not bad — much of the time. It usually lets us see ourselves through rose-colored glasses, which keeps depression at bay. Removing those glasses increases the risk for depression. Wearing them all the time leads to illusory optimism and excessive risk taking. If we use the cool system to monitor and correct the distortions from the rose-colored glasses, perhaps we can avoid hubris and some of the hazards of overconfidence. We can benefit from the psychological immunities that protect us from feeling terrible, that help us develop a sense of agency and efficacy in our lives, and that enable optimistic expectations that in turn reduce stress and sustain mental and physical health. I look at how these processes play out, and how they can be harnessed by the cool system to enhance our lives.

Western conceptions of traits and human nature have long assumed that self-control and the ability to delay gratification characterize individuals consistently, and will be reflected in their behavior across many different situations and contexts.

This is why much shock and surprise are expressed in the media each time the world learns about another famous leader, celebrity, or pillar of society whose hidden life has been exposed, revealing what appears to be a massive failure of judgment and self-control. These people must be able to wait for their marshmallows and to delay gratification in many situations — otherwise they could not have achieved their remarkable success. Why then do smart people so often act stupidly, managing to unravel the lives they diligently constructed? What trips them up? To understand this, I look closely at what people really do, not just what they say, across different situations and over time. There is consistency in the expression of traits like conscientiousness, honesty, aggression, and sociability. But it is consistency contextualized within specific types of situations: Henry is always conscientious, *if* at work but not *if* at home; Liz is warm and friendly, *if* with close friends but not *if* at a big party; the governor is trustworthy *if* dealing with his state's budget, but not *if* surrounded by attractive assistants. Consequently, we have to look at the particular situations in which people are or are not conscientious, sociable, and so on if we want to understand and predict what they are likely to do in the future.

The past few decades of discoveries, especially in social cognitive neuroscience, genetics, and developmental science, have opened new windows into how the mind and brain work — and they have done so in ways that make self-control, cognitive reappraisal, and emotion regulation central players in the story of who we are. It has even turned young philosophers into experimentalists, testing fresh ideas about human nature in the real world — not only who we are but also what we can become. The situations and skills that allow us to have agency, exert control, and make informed choices are far from limitless. They are hedged by the formidable constraints of living in a largely unpredictable world in which good and bad luck, as well as our social and biological histories and our current environments and relationships, make their contributions and limit our options. Yet the self-control skills we develop can make a substantial difference if we use the cool system flexibly and with discrimination, refusing to let it become rigid or allowing it to squeeze the pleasure and vitality out of the hot system.

What drives the cool system is the prefrontal cortex, as I emphasized in Part I. It enables the attention control, imagination, planning, and thinking needed to solve problems and exercise effort and self-control in pursuit of long-term goals that allow preschoolers to wait for treats. The same strategies work over the course of life — it's just the temptations that change. How and why these strategies work, and how they might make a difference to your life, is the stuff of Part II.

8

THE ENGINE OF SUCCESS: "I THINK I CAN!"

PART I LEFT US with a critical unanswered question: how do we make sense of the correlations between the number of seconds preschoolers wait for their bigger treats and how their lives turn out? This chapter unpacks the connections to show how the ability to voluntarily exercise self-restraint in pursuit of a hot goal early in life provides children with a powerful advantage that can help them succeed and maximize their potential throughout their lives. While self-control ability is an essential ingredient for constructing a good life, it does not function in isolation: the engine of success is fueled by additional resources that protect against the negative effects of stress and provide a foundation that can be cultivated and nurtured. In this chapter, I look at these resources and how they work. I begin with George, whose young life illustrates the research.

A SAVED LIFE: GEORGE

Far from the privileged world of the Surprise Room in Stanford University's Bing Nursery School, George Ramirez (his real name, used with his permission) grew up in one of the most impoverished areas of New York's South Bronx. George was born in 1993 in Ecuador, where his father worked in a bank and his mother was a librarian. When he was five years old, the economy "went bad," and he, his older sister, and his parents immigrated to the South Bronx with little money. The family lived together in one room, and George was enrolled in Public School 156 four blocks away. I met him when he was nineteen and we talked about his first experiences there:

> I spoke no English. They put me into a bilingual class. My teacher was really nice. It was just a mess. People running everywhere, screaming, adults screaming, total confusion, pushed around, terrified, no instruction....I got into a few fights and was constantly surrounded by adults who directly and indirectly told me and my classmates I was getting nowhere. "Why do I even bother trying?" I remember my second-grade teacher yelling over my rowdy class. "It's not like you'll actually make anything of yourselves."...And it stayed that way for four years.

When George was nine years old, his family won the lottery that let him enroll at KIPP, the Knowledge Is Power Program, a charter school that I describe further in Part III and that, he says, "saved my life."

I met George in 2013, when he came back to volunteer at

KIPP as an alumnus, helping young students make the most of their experience. Discussing the public school he had attended, which was located in the same building three floors below, he commented, "I'm sure they're trying but it still feels the same." Noise could occasionally be heard in the KIPP hallways from the public school classrooms below. I did not visit those floors, but George's descriptions fit my own impressions of the nearby South Bronx public middle school, where my students and I had done research a few years earlier.

I asked George how KIPP had "saved" him:

> The first time I came to KIPP is the first time anyone believed in me. My parents encouraged me but as parents without knowledge; KIPP encouraged me with knowledge and gave me "We believe in you, so let's do this! Here are the resources." The long hours, the orchestra, the focus on character and college preparation, the "tough love," and the positive expectations. "*All* of you will go to college!" It's showing that you care by being very, very honest. If you make a mistake and do something that doesn't make you smart, they show you what you need to do, and you know they do it because they care.

George believes that the most important way KIPP changed him was by making it clear that there were consequences for his behavior:

> Explicit expectations for the first time in my life that there *are* consequences. I had never been at a place

> where people told me what they wanted out of me — without screaming. And what they wanted was for my own good, and everyone else's. Plus lots and lots of positive reinforcements for doing well, and for everything good I did. When you do the right thing, the right things happen. When you do the bad, wrong thing, the bad things happen.

George learned quickly about the consequences: "In one year I generalized this to life outside the school. If I'm polite to others, they're polite to me. It usually, but not always, works in the real world. Soon you generalize the rules of 'consequences to my actions' from here to everywhere."

When he came to KIPP in 2003, George wasn't a bad student, but he had a short temper, was rude, and was really quiet. "Whenever I didn't get what I wanted I'd really get upset, bad temper, no self-control, found everything amusing at the wrong time — laughed when people were inappropriate." He got into trouble on his first day at KIPP and was shocked when he was told to stand at the back of the classroom for rolling his eyes at his math teacher. He was even more surprised when he was assigned homework and it was thoroughly checked the next day, something he says he had never experienced in his public school.

George attributes much of his school success to hard work. His days at KIPP were long: he arrived at school at 7:45 a.m. and stayed until 5 p.m. or sometimes even 10 p.m. Once he got home he faced hours of homework, and he went to school on Saturdays and for a couple of weeks every summer. My grandmother would have liked him. As she used to tell everybody, the magical ingredient in making an exceptionally successful life is what she called *sitzfleisch.* She meant sitting on your behind and putting in the huge effort needed to get the job done. My grandmother's focus on sitzfleisch was echoed a lifetime later by Bruce Springsteen, the rock musician, songwriter, and performer, who seems to personify the qualities that underlie a vibrant, fully realized life. Born in 1949, Springsteen has continued to perform brilliantly into his sixties, exhilarating his adoring crowds, and has been the subject of historical exhibitions at the National Constitution Center and at the Rock and Roll Hall of Fame Museum. Asked before a performance what he thought the inner qualities were that made him the artist and performer he has become, he said, "I probably worked harder than anybody else I saw."

At the time of this writing, George was doing extremely well, working toward his bachelor's degree on a full scholarship at Yale University. I asked him where he thought he would be if he had not won the lottery that allowed him to transfer into KIPP. "Without KIPP I absolutely would be hanging in the streets, looking for a job," he replied. What was at the root of his transformation from feeling totally adrift at age nine to becoming a successful Yale undergraduate? He said, "Learning to have self-control, being honest, being kind to my teammates, being polite, never settling for what I have, and asking the big questions were all things that led to my success at KIPP and in life."

When I'm asked "Isn't the future already prewired and visible in the child from the very start? Isn't that what the marshmallow studies tell us?," George's life is my answer. He surely had much good prewiring and potential, but as he emphasizes, there is no way his life would be unfolding as it is if he had not been "saved" by KIPP. Whatever his genetics, he was not on his way to Yale. The KIPP experience and the support, knowledge, resources, and opportunities its people provided allowed George to go from being adrift to launching a fulfilling life.

George would not have benefited so much from this kind of program if he had not worked so hard from age nine on. It's not just George, and it's not just the world of mentors, models, resources, and opportunities that KIPP gave him. It is both nature and nurture, not in opposition but influencing each other reciprocally as their boundaries blur. How a person interacts with that world of opportunities and constraints drives the life that unfolds. The sobering thought is that it took winning a lottery to give George his chance.

When George arrived in the South Bronx at age five, in a new country with a new language, perhaps he was ready to develop an "I think I can!" orientation to his life. His first public school was supposed to help identify, nurture, and educate his talents and prepare him for further learning. Instead, it plunged him into a confusing "jungle," as George called it. Fortunately, he attributed his confusion and sense of being adrift to the school and the circumstances, not to himself. Even after four years of chaos, he still felt "I was not a bad student." He recognized that he had a short temper and was rude but did not seem to question his ability to learn.

EXECUTIVE FUNCTION: THE MASTERY SKILLS

George Ramirez did not take the Marshmallow Test as a four-year-old, but his journey from the South Bronx to Yale University illustrates that he had the cognitive skills the test measures:

his cool system was functioning well, allowing him to control impulsive tendencies and hot reactions when he was motivated to do so. He managed this by using a part of the cool system critical for self-control called executive function (EF). These are the cognitive skills that let us exert deliberate, conscious control of thoughts, impulses, actions, and emotions. EF gives us the freedom to inhibit and cool impulsive urges, and to think and deploy attention flexibly in ways that let us pursue and reach our goals. This set of skills and neural mechanisms is essential for constructing a successful life.

The preschoolers and kindergartners who waited for their marshmallows or cookies showed us what EF is and how it works, letting us see what they had to do to restrain themselves from ringing the bell or nibbling on their temptations. Recall Inez, for example, who peeked at her cookies, reminded herself of her goal, and then quickly self-distracted to reduce the temptation. She began to invent little self-teasing games to amuse herself. She toyed with the bell while being careful not to ring it, shushed herself with her fingers pressed in front of her lips as if to say "No, no" to herself, beamed with pleasure and self-congratulation at her performance, and kept it up until she reached her goal.

Each child who waited successfully had a distinctive methodology for self-control, but they all shared three features of EF: First, they had to remember and actively keep in mind their chosen goal and the contingency ("If I eat the one now, I don't get the two later"). Second, they had to monitor their progress toward their goal and make the necessary corrections by shifting their attention and cognitions flexibly between goal-oriented thoughts and temptation-reducing techniques. Third, they had to inhibit impulsive responses — like thinking about how appealing the temptations were or reaching out to touch them — that would prevent them from attaining their goal. Cognitive scientists can now see these three processes play out in the brain when people who try to resist temptations are imaged in fMRI scanners, revealing the attention-control network in the prefrontal cortex that enables these remarkable human feats.

EF allows planning, problem solving, and mental flexibility, and it is essential for verbal reasoning and school success. Children who have well-developed EF can inhibit impulsive responses, keep instructions in mind, and control their attention when they pursue their goals. It is not a surprise that these children do better during the preschool years on math, language, and literacy tests than peers with weaker EF.

As executive function develops, so do the brain regions that enable these skills, mostly in the prefrontal cortex. As Michael Posner and Mary Rothbart showed in 2006, the circuits involved in EF are closely interconnected with more primitive brain structures that regulate the developing child's reactions to stress and threat in the hot system. These close neural interconnections are why long-term exposure to threat and stress undermines the development of strong EF. When the hot system takes over, the cool system suffers, and so does the child. On the flip side, however, well-developed EF helps regulate negative emotions and reduce stress.

If EF is severely impaired, our prospects are limited. Without EF, it becomes impossible to control emotions appropriately and inhibit interfering impulsive responses. Children need EF to resist temptations beyond marshmallows — for example, when they have to stop themselves from striking another child who accidentally spilled juice on their new shoes. Children who lack EF have a hard time following directions and are prone to aggressive confrontations with both adults and peers, setting them up for trouble at school. Even those who are predisposed to aggressive "acting out" are not as fiercely aggressive if they are able to self-distract to cool themselves down (see Chapter 15). These skills help kids not only delay gratification but also control their anger and hot negative impulses.

Preschoolers need EF when they face hot tasks such as the Marshmallow Test or when Mom leaves the room. But outwardly cool tasks may require EF as well. For example, an ostensibly cool task like learning arithmetic in school can easily become hot when fear of failure and performance anxiety activate the hot system and attenuate the cool system, escalating stress and undermining learning. And what is hot for one person may be cool for another. People with good EF for one type of challenge may find other challenges more difficult. Some children are excellent in the classroom setting, for example, but have uncontrolled tempers when interpersonal situations trigger their hot spots. Others have the reverse pattern: they can be interpersonally cool but become stressed and lack cognitive control in school situations that require concentration and focused effort.

Children who develop good EF in preschool are better prepared for dealing with stress and conflicts triggered by hot visceral temptations. The same skills routinely help them when they are learning how to read, write, and do math. On the other hand, if EF does not develop well in preschoolers, which happens much too often, those children are at increased risk for ADHD and a variety of other learning and emotional problems throughout their school years.

EF, IMAGINATION, EMPATHY—WINDOWS INTO THE MINDS OF OTHERS

Because EF requires us to exert cognitive control over our thoughts and feelings, it is easy to think that it is the antithesis of creative and imaginative processes. But in fact, it appears to be an essential ingredient for the development of imagination and creative activities, including pretend play early in life. EF allows us to get beyond the immediate situation and the here and now, to think and fantasize "outside the box" or imagine the impossible. By facilitating imagination, EF in turn enhances the development of flexible and adaptive self-control. Likewise, EF is strongly linked to the ability to understand the mind and feelings of others, and helps children develop a "theory of mind" for inferring the intentions and anticipating the reactions of people with whom they interact. EF allows us to understand and take into account the feelings, motivations, and actions of others and recognize that their perceptions and reactions may be quite different from ours. It helps us grasp what others may think or intend and lets us empathize with what they are experiencing.

Our theory of mind may be related to "mirror neurons," which Giacomo Rizzolatti discovered in monkeys. Although we share these neurons with monkeys, we are much better at empathy than they are, and that difference is an important part of what makes us human. The role of human mirror neurons is still debated, but they seem to be part of the neural structures that allow us to experience a milder version of what others are thinking and feeling. These mirrors in our minds make us smile when someone friendly smiles at us. They make us afraid when others are frightened and cause us to feel pain or joy when others do. As Rizzolatti put it, these mirrors let us "grasp the minds of others not through conceptual reasoning but through direct simulation. By feeling, not by thinking." They are fundamental for our functioning and survival as interdependent social creatures living together in society.

THE ENVIABLE BELIEFS

If EF is well developed early in life, children have a better chance to construct the lives they want. They have a foundation for building interconnected beliefs about themselves that should rank high on our wish list for those we love: a sense of personal control or mastery reflected in an "I think I can!" mind-set, and optimistic expectations about the future. It is important to understand that these enviable "resources" are the individual's *beliefs* about the self, not external evaluations or objective tests of achievement or competence. Just as the negative effects of stress depend on the individual's *perceived* stress, and the impact of temptations depends on how they are appraised and mentally represented, the potential health benefits of our abilities, achievements, and prospects depend on how we interpret and evaluate them. Think of people you know who are highly competent but sabotage themselves with their own negative self-evaluations and paralyzing self-doubts. Beliefs about the self are correlated with objective measures of competence and mastery, but far from perfectly.

The impressive evidence about the importance of these beliefs for successful coping, both psychologically and biologically, keeps growing. Shelley Taylor, the founder of the field of health psychology and a professor at the University of California in Los Angeles, and her team have shown that a sense of mastery and optimistic expectations buffer the deleterious effects of stress and predict many desirable neurophysiological and psychological health-related outcomes. As Taylor and her colleagues reported in 2011 in the *Proceedings of the National Academy of Sciences*, each belief has a substantial genetic component but is also open to modification and influence by environmental conditions. Given the importance of these beliefs for the quality and length of life, I next consider each more closely.

MASTERY: PERCEIVED CONTROL

"Mastery" is the belief that you can be an active agent in determining your own behavior, that you are able to change, grow, learn, and master new challenges. It's the "I think I can!" belief that George Ramirez says KIPP taught him and that turned his life around. I first saw its importance when I was a trainee in the clinical psychology doctoral program at Ohio State University and watched my mentor, George A. Kelly, work with a very distressed young woman. "Theresa" was becoming increasingly upset and anxious, feeling that she could no longer manage her life. In the third therapy session, her agitation seemed to peak, as she exclaimed tearfully that she was afraid she was losing it and begged Dr. Kelly to answer her question: "Am I falling apart?"

Kelly slowly took off his glasses, brought his face close to hers, stared straight into her eyes, and asked, "Would you like to?"

Theresa was stunned. She seemed immensely relieved, as if a huge burden had dropped from her shoulders. It had not occurred to her that it might be within her power to change what she felt. "Falling apart" was suddenly an option, not her inevitable fate. She did not have to be the passive victim of her biography, witnessing her life unravel. This was her "Eureka!" moment; it started her exploration of alternative and more constructive ways of thinking about herself and opened courses of action she had not considered because she had thought them impossible.

Carol Dweck, my colleague for many years at Columbia University (she is now at Stanford University), has become one of the most forceful and effective modern voices in psychology on the topic of perceived control and mastery beliefs. Her work, summarized in her 2006 book, *Mindset*, shows how people's personal theories about how much they can control, change, and learn — and how much they can improve what they do, experience, and make of themselves — influence what they actually can achieve and become. Dweck and her colleagues demonstrate that these personal theories about the malleability versus fixed state of one's characteristics matter greatly, whether in regard to self-control and willpower, intelligence, mental state, or personality. These theories change how we evaluate our performance, how we judge ourselves and other people, and how the social world in turn responds to us.

Beginning early in life, some children see their intelligence, their ability to control the world around them, their sociability, and other characteristics not as fixed essences that they are either stuck with or blessed with from birth on, but more flexibly, like muscles or cognitive skills that they can build and develop. Dweck calls these children "incremental growth-minded theorists." Others, the "entity theorists," see their abilities as frozen from birth at some fixed level that they cannot change: smart or dumb, good or bad, powerful or helpless. Happily, Dweck's work goes beyond showing the importance of these mind-sets. Her research also makes clear that mind-sets are open to change and illustrates many ways to rethink and modify them.

Dweck shows that children stuck with an entity theory about their own abilities tend to have a particularly difficult time when schoolwork becomes increasingly challenging. This is especially striking in the transition from elementary school to junior high school in the United States, when many schools suddenly grade competitively rather than reassuringly. The school experience shifts from fun and easy to tough and demanding, with long and hard homework and competitive peers. Dweck found that, under high pressure and the threat of failure in the new school environment, the students who viewed their abilities as fixed — the entity theorists — soon began to get lower grades, doing progressively worse over the two years of junior high school. The students with a growth mind-set, in contrast, kept getting better and better grades over those two years. When the two groups began junior high school, their past school records could not be distinguished. By the end, there was an obvious gap.

Students with the fixed mind-set justified the difficulties they were having with the tough new school demands by maligning their own abilities — "I suck in math" or "I'm the stupidest" — or by blaming their teachers: "The teacher is on crack." Those with the growth mind-set also sometimes felt overwhelmed by the new demands, but they responded by digging in, figuring out what it took to master the new situation, and doing it.

At the preschool level, the belief that "I think I can!" is captured in the classic story for young children *The Little Engine That Could.* A train filled with toys and treats for little boys and girls gets stuck as it is trying to get up the last steep mountain. A shiny new passenger train and a strong freight train and an old tired train pass by, all refusing to help. At last, the kind Little Blue Engine arrives. It struggles and struggles, chanting its growth-minded theory "I think I can — I think I can — I think I can," until it ultimately prevails and delivers the gifts to the children waiting on the other side of the mountain.

At Stanford in 1974, my students and I developed a scale to assess how preschoolers perceived the causes of their own behavior: did they see themselves as the agents for the good things that happened, or did they credit external factors? And did these differences in causal attribution link to their self-control and how they were developing? In order to measure where they were on this scale of perceived "internal versus external control of behavior," we asked them questions like these:

> *When you draw a whole picture without breaking your crayon, is that because you were very careful? Or because it was a good crayon?*
>
> *When somebody brings you a present, is that because you are a good girl (boy)? Or because they like to give people presents?*

We then looked at how the children's answers to these questions were connected to their behavior, which was also assessed in other situations that required self-control. The bottom line from these studies was that even preschool children's belief that they could control outcomes by their own behavior was significantly linked to how hard they tried, how long they persisted, and how successful they were at self-control. The more they saw themselves as the causes of positive outcomes, the more likely they were to delay gratification on the Marshmallow Test, to control their impulsive tendencies, and to persist in diverse situations in which their own behavior would be instrumental in reaching the desired outcome. They believed they could do it, and they did.

The child's self-perception as someone who *can* — who can exert effort, persist, and be the causal agent for positive outcomes — is nourished by the self-control skills that help them succeed. One could see this in the pride that some preschoolers expressed in the Surprise Room at Stanford when, instead of just eating their treats for which they had successfully waited, they chose to bag them to take home, eager to show their parents what they had earned. The more effectively children can wait and work for their bigger treats early in life, and the better the cognitive and emotional skills that enable these triumphs, the more they grow their sense of "Yes, I can!" and ready themselves for new and greater challenges. In time, the mastery experiences and new skills they acquire — like learning to play the violin, or build Lego empires, or invent new computer applications — become intrinsic rewards in which the activity itself is satisfying. The children's sense of efficacy and agency becomes grounded in their experiences of success and leads to reality-based optimistic expectations and aspirations, each success increasing the chances for the next.

OPTIMISM: EXPECTATIONS OF SUCCESS

Optimism is an inclination to anticipate the best possible outcome. Psychologists define it as the extent to which individuals have favorable expectations for their future. These are expectations of what they really believe will happen — more like faith than just hope — and they are closely linked with the "I think I can!" mind-set. The positive consequences of optimism are dazzling, and would be hard to believe if they were not so well supported by research. For example, Shelley Taylor and her colleagues showed that optimists cope more effectively with stress and are better protected against its adverse effects. They take more steps to protect their health and future well-being, generally stay healthier, and are less likely to become depressed compared with those who are low in optimism. Psychologist Charles Carver and his colleagues showed that when optimists have coronary bypass surgery, they recover more quickly than pessimists do. The list of benefits goes on and on. In short, optimism is a blessing to be wished for, as long as it reasonably connects to reality.

To appreciate optimism and see why and how it works for those who have it, consider its opposite: pessimism. Pessimism is a tendency to focus on the negative, expect the worst, or make the gloomiest interpretation possible. Show a depressed pessimist a phrase like "I really hate" followed by a blank space in which to insert the first thoughts that come to mind, and he or she is likely to insert "me" or "the way I look" or "the way I talk." Extreme pessimists feel helpless, depressed, and unable to control their lives. They attribute the bad things that happen to them to their own stable negative qualities, rather than being open to more situational and less self-condemning explanations of what went wrong. They fail a test and think, "I'm incompetent," even if the test is not a valid measure of anything important. Kinder explanations — whether regarding the test itself (confusing instructions, ambiguous multiple choice options, excessive time pressure) or personal problems (an upset stomach) — don't come to mind for pessimists, even if they happen to be true.

Early in life, if this pessimistic explanatory style is extreme, it can be bad news for the future and can become a formula for developing serious depression. At the University of Pennsylvania, Christopher Peterson and Martin Seligman asked healthy 25-year-old college graduates to describe some of their difficult personal experiences and then examined how they explained them. The pessimists believed that things would never improve ("It won't ever be over for me") and generalized broadly beyond each event to reach gloomy conclusions about diverse aspects of their lives, all of which they considered their fault. Follow-up health examinations and measures of illness for all participants during the first 20 years after college showed no significant differences in their health. Between the ages of 45 and 60, however, those who had been more pessimistic at age 25 were more likely to be ill. Researchers also analyzed newspaper interviews with ballplayers from the Baseball Hall of Fame published throughout the first half of the last century. The interviews quoted the players' explanations as they talked about how and why they won or lost games. These players were all outstanding enough to be in the Hall of Fame, but those who saw their losses as due to their personal failings, and attributed their wins to momentary external causes (e.g., "The wind was right that afternoon"), tended not to live as long as those who took credit for their successes.

Seligman has led much of the research on optimistic versus pessimistic explanatory styles. He proposed that optimists differ from pessimists in how they perceive and explain their success and failure. When optimists fail, they think they can succeed the next time if they change their behavior or the situation appropriately. They use a rejection experience, failed job application, bad investment, or poor test result to figure out what they need to do to improve their chances on the next attempt. They then craft alternative plans and find other ways to reach their important goals, or seek needed advice until they can develop a better strategy. While the optimists deal with failure constructively, the pessimists use the same experience to confirm their gloomy expectations, believing it's their fault, and they try to avoid thinking about it, assuming there is nothing they can do. Seligman says, "College entrance exams measure talent, while explanatory style tells you who gives up. It is the combination of reasonable talent and the ability to keep going in the face of defeat that leads to success.... What you need to know about someone is whether they will keep going when things get frustrating."

This is an equally apt description of the preschool kids who continue to wait during the Marshmallow Test. Seconds of waiting time not only measure their delay ability; how long they waited also tells us how much grit the children have, or how persistent they are as the frustration of the delay and the effort needed to stick with it keep escalating. Because optimists have higher overall expectations of success, they are more willing to delay gratification, even when it is difficult to do so. Unless children expect to succeed and get those marshmallows later, when the experimenter comes back, there is no reason for them to try to wait or work for them. Those who expect that they will be able to do whatever it takes to get their preferred rewards choose to wait and work for them; those who don't (or who don't trust the experimenter) take the immediately available smaller rewards by ringing the bell.

Ervin Staub escaped from communist Hungary as a young man, and in the early 1960s he became one of my first graduate students at Stanford as well as a lifelong friend. Together we conducted experiments at Stanford to see how expectations about success influence self-control and the willingness to work and wait for delayed rewards. We found that 14-year-old eighth-grade boys who generally expected to succeed even before they saw the specific task they would have to perform chose to do cognitive tasks for which the larger but delayed reward was contingent on successful performance, not just on waiting. This was rather than settling for the smaller but immediate reward, and they chose this option almost twice as often as those with low success expectations. The boys with high expectations for success approached new tasks more confidently, as if they had already succeeded at them. They wanted to "go for it," and they were willing to risk failure because they did not believe they would fail. Their expectations were more than fantasies: they were based on their history of previous successful experiences. Their successes fed the positive expectancies, which in turn encouraged behaviors and mind-sets that increased their chances for further success. All of which makes optimists smile even more.

The findings also showed that those who started with low generalized expectations began as if they had already failed at the task. But these boys did respond positively when they actually succeeded at it, and their new success experiences significantly raised their expectations for future success. Our broad expectations for success or failure crucially impact how we approach new tasks, but our specific expectations are responsive to change when we see that we can actually succeed. The message is clear: optimists in general are better off than pessimists, but even pessimists raise their expectations when they see that they can succeed.

VIRTUOUS AND VICIOUS CYCLES

In sum, the successes and mastery experiences children have early in life increase how willing and able they become to pursue goals with persistence, develop optimistic expectations for success, and cope with the frustrations, failures, and temptations that are inevitable as they grow up. Their developing sense of control and agency and optimistic expectations become key links — the active ingredients — in the story that connects the seconds of preschool waiting time for a couple of marshmallows to the diverse positive outcomes we see as their lives evolve. And their ability to inhibit impulsive responses that could jeopardize their building of relationships allows them to develop mutually supportive, caring friendships with people who respect and value them.

This chapter described a virtuous cycle of growth to hope for and nurture in our children. It contrasts with the vicious cycle faced by children who persistently lack basic self-control skills, feel out of control, are pessimistic about their own abilities, and struggle to maintain a sense of self-worth. Without adequate self-control skills, optimistic expectations, success experiences, and the help and support of others, children may remain largely controlled by their hot system, more likely to fail in their earliest efforts at mastery, and prone to develop feelings and mind-sets of helplessness rather than hopefulness, as their choices and options shrink.

9

YOUR FUTURE SELF

THE ANT IN AESOP'S fable instinctively knows what it has to do to prepare for the future, and when summer comes it drags off the food it will need for the winter. But we don't have the ant's instincts, and evolution has not yet adapted our brain for dealing concretely with the distant future. We easily become anxious about frightening events that are imminent but rarely visualize the future in vivid, hot terms. Those rose-colored glasses and the feel-good psychological immune system protect most of us from dwelling on such anxieties. They allow us to avoid focusing on terrifying prospects like cancer, impoverishment, lonely old age, and ill health, and if these anxieties do become vivid, most of us soon self-distract.

In this way, we avoid the anxiety that Freud found in his patients, and that the painter Edvard Munch depicted in *The Scream.* An icon of anxiety in the modern world, the painting shows a terrified person trembling on a bridge in ominous surroundings, hands cupped against the ears, eyes wide open staring at us from a horror-struck face. Our defenses protect us from lingering too long on such an image, but they also make it unlikely that we'll behave like provident ants rather than self-indulgent grasshoppers. Consequently, people continue taking all sorts of risks, like eating too much and smoking and drinking too heavily, ignoring the long-term consequences that are far off, uncertain, and easily discounted. The vast majority of Americans arrive at retirement age with funds completely insufficient to maintain anything remotely like the lifestyle to which they have become accustomed. The problem begins with how we naturally think about the future self and how that future self is represented in the brain.

MULTIPLE SELVES

Shakespeare's "Seven Ages of Man" famously captured the multiple selves experienced in the course of life:

All the world's a stage,
And all the men and women merely players;
They have their exits and entrances;
And one man in his time plays many parts,
His acts being seven ages.

Shakespeare begins with the infant, "Mewling and puking in the nurse's arms," and, after describing our young and middle ages, moves on to old age:

The sixth age shifts
Into the lean and slipper'd pantaloon,

With spectacles on nose and pouch on side,
His youthful hose, well saved, a world too wide
For his shrunk shank; and his big manly voice,
Turning again toward childish treble, pipes
And whistles in his sound. Last scene of all,
That ends this strange eventful history,
Is second childishness and mere oblivion,
Sans teeth, sans eyes, sans taste, sans everything.

The human body changes completely as we age, but does the self we experience change too? What happens when you travel through time in your imagination and think about yourself in the future? Take a close look at the pairs of circles below that move from no overlap between your current and future self to almost complete overlap. Pick the pair that best depicts the degree to which you feel yourself similar and connected to the future self that you expect to be ten years from now, and mark it.

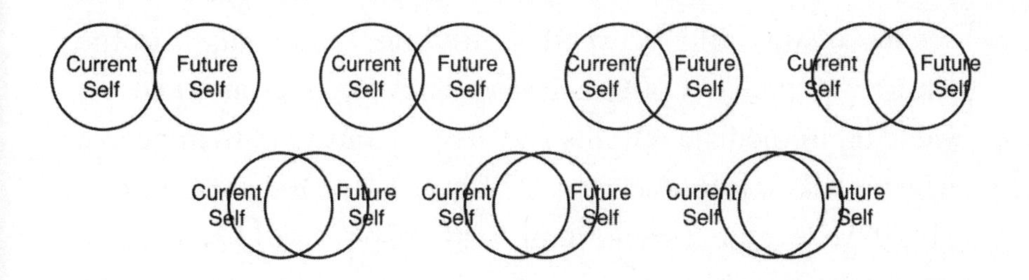

Now imagine that you have agreed to let your brain activity be imaged while you are inside an fMRI machine. Your head is deep in the machine, and you are getting used to the confined space as the instructions come through the speaker: "Please think about yourself." When you do, a distinctive pattern of brain activity will become visible in the midfront area of your cerebral cortex, which we'll call the self pattern. Next, the instructions ask you to think about a stranger, and in the same area of your cerebral cortex, a distinctively different pattern becomes activated, the stranger pattern. Finally, you are asked to "please think about yourself ten years from now."

Hal Hershfield, now at New York University, and his colleagues conducted this study with Stanford University undergraduates in 2009. They found that we differ not only in what we *feel* when we imagine our future selves but in our brain activity, depending on how closely we connect our self-perceptions and identity in the present to the selves we become. For many people, the pattern that becomes activated in the brain for the future self is more like the one for the stranger than like the one for the self. But there are individual differences, showing that some people are more emotionally in touch and identified with their future self, while for others that older person might as well be somebody else.

How much overlap is there in the circles you selected? If you see more continuity between yourself now and yourself in the future, you probably put more value on delayed rewards and less value on immediate rewards and are less impatient than people who view their future selves as strangers. As the researchers point out, if we feel greater continuity with who we will become, we might also be willing to sacrifice more of our present pleasures for the sake of that future self.

The same group of researchers also looked at financial decision making in adults (mean age of 54 years) living in the San Francisco Bay Area. The participants who perceived greater rather than less overlap between their present self and their future self not only had a stronger preference for delayed larger rewards rather than immediate smaller rewards, but also accumulated more real-world assets (net worth from all sources) over time. When I read Hershfield's research I reminded myself to recheck the status of my retirement plan.

MONEY NOW VERSUS 401(K) FOR THE FUTURE

If Adam and Eve had managed to cool the hot temptations they faced, they would have been able to hold on to their garden longer. If they had wanted to prepare for what they might face in the future, they would have had to vividly imagine themselves in it, which they could not do. The kids in the Surprise Room had to cool down to resist taking their one marshmallow. Decades later, when making their 401(k) choices, they have to imagine themselves in old age, not abstractly but concretely, in order to heat the scene emotionally as if they were already there. While they're still young, they have to linger on that future self — at least long enough to check the box on the 401(k) retirement form that initiates the retirement plan.

Just as preschoolers' willingness and ability to wait for two marshmallows depend on how they mentally represent the treats, the young adult's ability to connect with the self expected years in the future depends on how that distant self is mentally represented. To explore this, Hershfield and his colleagues set out in one study to show a group of college students vivid representations of their retirement-aged selves while they were making financial decisions. As a first step, the researchers asked each participant for a photo of him- or herself and then created an avatar (or digital representation) out of that photo. For some of the participants, the avatar was of themselves at their present age; for others, the avatar was made to look older, representing the person at approximately age sixty-eight. Participants used a slider scale with an arrow to indicate what percent of their hypothetical paycheck they would allocate toward their 401(k) retirement account. As they moved the arrow to the left, they increased the percent they would get as take-home pay, and as they moved the arrow to the right, they increased the percent going into their retirement fund.

Participants saw either the avatar of their current self (displayed on the left side of the slider) *or* the avatar of their future self (displayed on the right side of the slider). Would calling attention to the self they would become in the future influence the present self to share current income? Indeed, those who saw their future self indicated that they would save 30 percent more relative to those who saw their current self.

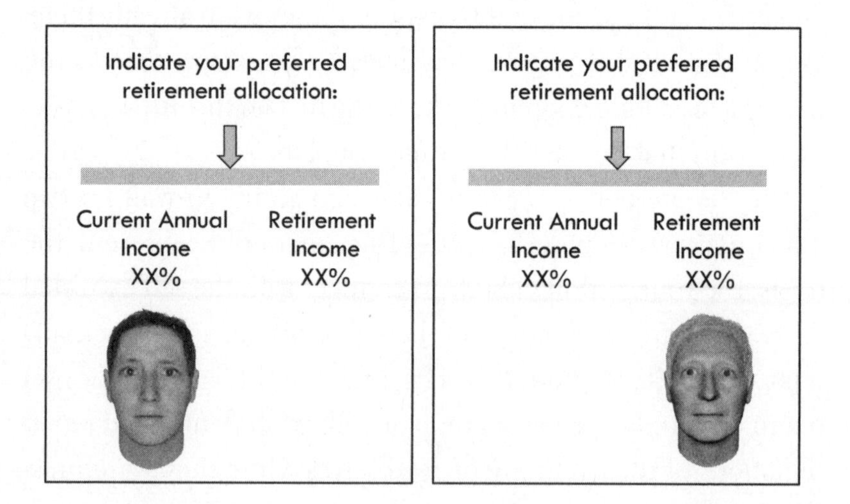

The idea guiding this research is that the more emotionally connected you become to your future self, the more you will incorporate it into your present self-conception and budget, ready to share more generously from what you currently give yourself to what you allocate to yourself in the future. Hershfield and other researchers are continuing to explore whether savings for later years, not just in hypothetical lab situations but in real life — specifically in 401(k) retirement plans — might be substantially increased by enhancing the saver's identification with the future self.

BECOMING ETHICAL FOR THE FUTURE SELF?

If you feel closely connected to your future self, will you be more likely to take into account how your present actions will impact how you feel later — and not just when it comes to budgets and retirement planning? Specifically, will individuals who experience close continuity with their future selves be less likely to make the unethical decisions that are all too common in everyday life? It's a timely question, given that FBI statistics on white-collar crime, first gathered in 1940, indicate that the rate of such crime had tripled by 2009 — a problem made salient by the rash of financial scandals exposed during the 2008 financial crisis, including Bernard Madoff's monumental Ponzi scheme. In 2012, Hershfield and his colleagues asked that question in five online studies using women and men ranging in age from 18 to 72 years. The researchers inquired about their willingness to endorse unethical but profitable business decisions and their moral judgments about the acceptability of lies and bribes in business contexts. For example, how likely would the individual (always anonymous) be to market a profitable food product with known health hazards or to endorse a financially lucrative but environmentally risky mining operation that could lead to a large bonus? Across the five studies, people who felt more disconnected from their future selves — measured by little overlap in their circles for the present self and the future self — were more apt to tolerate unethical business decisions.

The researchers also prompted some participants to think about their future selves while others were primed to think about the future world. Projecting the self into the future, compared with simply thinking about the future, reduced tolerance for unethical behavior. Those who felt closely connected to the future self thought more about the delayed long-term consequences of their actions, and it was this attention to future consequences that accounted for their unwillingness to make greedy, selfish decisions. It's a finding to recall before the hot system, oblivious to distant consequences and blind to ethics, faces its next tantalizing but immoral temptation.

10

BEYOND THE HERE AND NOW

SOME OF MY MORE memorable conversations with colleagues about how we think about the future have been at annual science conventions — not during meetings and research presentations, but when swapping personal stories late at night. Each of us had made decisions to accept invitations to give talks two or three years in the future, at unfamiliar, interesting-sounding places. One of my colleagues told the story of her 2008 invitation to give a special lecture in 2011 in an exotic small country thousands of miles away. When she got the invitation she asked herself, "Why go there?" She found many good reasons: The institute had some respected researchers working in her area; it was a new experience in an unfamiliar setting (a newspaper travel section called it "remote and beautiful"); she liked to travel to unusual places; her schedule for 2011 was still wide-open; and the organizers seemed very eager for her to come.

Two years later, as the time for the journey approached, her questions about the trip shifted from *why* she should go to *how* she would get there — and just what exactly she would have to do to make it happen. She had to arrange to change planes multiple times, flying with unfamiliar airlines that, on closer examination, had dubious safety records and bad histories for flight delays and cancellations. She had to renew her passport and get vaccinations, all while a seemingly endless series of unexpected events required her urgent attention. There was a complete shift in how my colleague thought about that trip from the time she gladly accepted the invitation to almost three years later, as her departure date approached. This surprised her, because as the future turned into the here and now, she wished she could call it off.

PSYCHOLOGICAL DISTANCE

Psychologists Yaacov Trope and Nira Liberman propose that when we imagine the future or think about the past we are traversing a single dimension: psychological distance. The distance can be in time (now versus in the future or the past), space (near versus far away), social (self versus strangers), and certainty (definite versus hypothetical). The greater the psychological distance, the more abstract and high-level information processing becomes, increasingly governed by the cool cognitive system. In my colleague's travel example, she had thought about the trip abstractly, without details and context in mind when it was far off in the future, and because it had seemed reasonable and made sense to her cool system, she had decided to go. As the psychological distance shrank, the more concrete, vivid, detailed, contextualized, and emotionally hot her information processing became, and the more she regretted her decision.

This shift in the level of information processing, from

abstract thinking about the future to concrete and vivid thinking in the present, affects what we feel and how we plan, evaluate, and make decisions. It helps explain why people make decisions about future events and commitments that they often regret — because when the future becomes the present, they find themselves facing a trip they don't want to take, an event they don't want to attend, a paper they don't want to write, or a family visit they don't want to make. The good news is that all will be well if we just wait and reflect for a bit after an event is over. The psychological immune system works hard so that we can look back and feel that the trip was worth taking, the event worth attending, the paper worth writing, and the family visit, on the whole, a good bonding experience.

To avoid the regret my colleague felt when her trip went from the hypothetical to the reality of packing and heading off to the airport, it might have helped if, before deciding to accept, she had imagined how it would play out as if it were happening now. If you want to decide how something (a new job, an exotic trip) will feel in the future, you might try to imagine yourself doing it in the present. Simulate the events as vividly as possible, in great detail, by essentially pre-living them. When my graduate students are fortunate enough to have more than one job offer and are tortured about their decision, I suggest that they imagine, as concretely as they can, living life in each job, one full day at a time, as if the job were happening now.

The research by Trope and his colleagues on how psychological distance influences us also speaks to why it is much easier to resist immediate temptations if we think about them in abstract, cool ways or as being far away in space and time. Such high-level, abstract thinking activates the cool system and attenuates the hot system. It reduces the automatic preference for immediate rewards, increases attention to future outcomes, strengthens intentions to exert self-control, and helps cool down hot temptations. Recall that when preschoolers push the treats far away, or turn around to avoid seeing them, or appraise their tempting treats at an abstract, cool level (imagine the marshmallows are just a picture; put a frame around them in your head), they are able to control themselves and wait much longer. But when they focus on the taste of the marshmallows and imagine appetitive, salient qualities in their mouths (yummy, chewy, sweet), delay becomes more difficult and they ring the bell.

INCREASING PSYCHOLOGICAL DISTANCE TO CONTROL CRAVING: CLOSE VERSUS FAR

Can people who are victimized by their dangerous cravings — whether for tobacco, alcohol, drugs, or fat-filled snack foods — cool them down by increasing their psychological distance? This question guided some experiments done by Kevin Ochsner and his team, with which I collaborated, at Columbia University. We wanted to try to help people control their cravings, beginning early in life. With that goal, we invited children and adolescents aged 6 to 18 to complete a task while their brains were being scanned with fMRI, and this gave us a glimpse into how they cognitively regulated their appetitive impulses. In the scanner, pictures of appetizing foods were quickly flashed in front of them in many different trials. In Hot, Close trials, we asked them to imagine that the food was close, right in front of them, and to focus on the hot, appetitive features like its taste and smell. In Cool, Far trials, we tried to activate their cool system by asking them to imagine that the food was far away and to focus on its cool, abstract, visual features (e.g., color and shape). The participants reported that they felt less craving during the Cool, Far trials than during the Hot, Close trials, and their brain images showed that reducing their craving decreased activity in the brain regions involved in appetitive desire.

Children in this study also took the Marshmallow Test, and their ability to control cravings for appetitive food was linked to their ability to delay on the test. Children who had not been able to wait as long for their treats experienced more craving during both Cool, Far and Hot, Close trials than those who could wait longer. And when they were trying to reduce their food cravings as their brains were being imaged in the scanner, the low-delay children also had less activation of the prefrontal cortex and more activation of the reward areas related to appetitive desire.

Hedy Kober, who was also working with Ochsner's team, led a similar study in which she showed photographs of cigarettes to induce cravings in heavy cigarette smokers. During each trial, while their brains were being imaged, participants were instructed to think of the item they saw by focusing either on its immediate, short-term effect "now" (e.g., "It will feel good") or on the "later" long-term consequences associated with consuming it (e.g., "I may get lung cancer"). When heavy cigarette smokers focused on the long-term consequences of cigarette smoking they significantly reduced their cigarette cravings.

Overall, we found that people can use simple cognitive strategies to regulate their cravings by shifting their time perspective from "now" to "later." You can turn such strategies into specific *If-Then* implementation plans like those discussed in Chapter 5, so that the temptation to smoke automatically activates your focus on the delayed negative consequences — making them hot and vivid enough to stifle the craving.

AN ADDICTED SMOKER'S SELF-GENERATED CURE

The Columbia studies tell us something about the mechanics of craving regulation that seems promising for real-life applications. But if it is really so simple, then why does the world still have such massive and costly problems with cravings? In research studies, the participants are volunteers who are willing to comply with the instructions and regulate their thinking accordingly, at least while in the laboratory. In the real world, it is of course a much more complicated story — as every addict who has tried to quit understands.

Carl Jung was reported to have said that people study what they are not good at themselves. This applies to me quite often. I am not a model of self-control — far from it — but I did have success in my struggle to overcome cigarette cravings. I tell my story here to show that even someone who is not broadly good at using this skill (and who often stresses his students and family with his impatience) can do it.

I began to tentatively puff on cigarettes as an adolescent, and it wasn't long before I became a habitual, heavy smoker. When the U.S. Surgeon General's report on tobacco risks came out in the early 1960s, my cool system briefly registered that smoking could result in serious long-term damage — and my hot system paid no attention. The cool system is rational, but it can also work closely with the hot system in the service of self-defense to cleverly rationalize whatever we do. In my case, it did an excellent job, letting me reappraise my smoking as just part of the lifestyle of an academic rather than a life-threatening addiction. I was a professor whose smoking allowed him to give less anxious, more thoughtful lectures. And everybody else was doing it anyway. So I kept chain-smoking while my cool system slept and my hot system enjoyed (and coughed).

One morning, I was in the shower with the water running before I realized that my lit pipe was still in my mouth. Denial was no longer possible: I was an addict, and I craved tobacco. At that point I was smoking about three packs of cigarettes a day and supplementing that with a pipe. My insight produced no behavior change; instead, it just increased my stress level. My cool system kept busy with other concerns.

Not long after that shower, I was walking through the halls of Stanford University's medical school and saw a sight that terrified me: a man was being wheeled down the hall, strapped to a gurney; his eyes were wide-open, staring at the ceiling, and his arms were stretched out at his sides. There were little green pen marks all over his exposed chest and on the top of his shaved head. A nurse explained that the patient had metastasized cancer and was being taken for another radiation session. The green marks were there to direct the radiation. I could not shake that vivid image of the consequences of my addiction from my mind. The Surgeon General's findings had finally penetrated my hot system and sounded the amygdala alarm.

Cigarettes were my continuous hot temptation, and I had to change them into something that disgusted me in order to cure my addiction. Whenever I felt a craving (at first this was often), I inhaled deeply from a large can filled with old, stale cigarette butts and pipe debris. The can had a concentrated nicotine odor intense enough to be nauseating. The textbooks call this aversive counterconditioning. I supplemented this step by deliberately reactivating that haunting image of the cancer patient to make the "later" consequences of smoking as hot, salient, and vivid for myself as possible. Perhaps just as important, I made a social contract with my three-year-old thumb-sucking daughter: she agreed to stop sucking her thumb and I vowed to stop sucking my pipe. I also made public commitments with my co-workers and students, contracting myself to quit cigarettes and no longer mooch them. It took a few weeks of struggle, but ultimately it worked. I still find that my hot system occasionally seats me at a table near a smoker in outdoor cafés. But after inhaling for a few moments, I almost always try to change my seat.

Visualizing yourself as a cancer patient being prepared for the next radiation treatment is anything but fun, and it makes your amygdala go wild with fear. But if your cool system wants to try it, it can be useful. This kind of visualization can be one step in conquering life-threatening addictions with deadly consequences that do not play out until much later, but whose prevention requires immediate self-control and delay of gratification. It demands doing exactly what does not come naturally: activating the hot system to make representations of the future more powerful than temptations of the present, and then using the cool system to cognitively reappraise the immediate temptations in ways that neutralize them or make them aversive within the hot system. At first it is effortful, but in time it becomes automatic.

PEEKING INTO YOUR DNA FUTURE

Taking the future into account in the decisions we make in the present requires imagining that future and predicting what it will feel like. Until late in the last century, attempts to predict the future were largely restricted to palm reading, tarot cards, astrology charts, soothsayers, and prophets. In Western history, future prediction dates from as early as the ancient Greek legends about the oracle at Delphi all the way to modern science fiction fantasies and fortune cookies. But in today's world, the cracking and decoding of the human genome lets us at last peek into our own DNA — which can be exciting to optimists but terrifying to pessimists. Soon you will likely be able to get a full report of the good and bad prospects that are waiting to unfold in your genes, for a fee no greater than that for a colonoscopy. That can be wonderful if you suffer from certain kinds of cancers or other diseases, offering the possibility of genetically targeted, individualized treatments based on your DNA that promise to overcome previously intractable medical problems. In such cases, getting the test may be an unambiguous choice. For most healthy people, however, whether or not to do such testing will be a wrenching decision — one in which the hot system will be overactive and the cool system will have a very hard time trying to help make a reasonable choice.

In the late 1990s, soon after the discovery of mutations in the BRCA1 and BRCA2 genes and their role in breast and ovarian cancer, many women faced a difficult decision. The choice to undergo testing for these mutations is particularly torturous because the psychological impact is likely to be both profound and unpredictable. Genetic testing can tell the population of people most vulnerable to these mutations that they either have a much greater chance of developing breast and/or ovarian cancer early in life or that they are in the lucky group unlikely to face such prospects. When the tests became available, many women were eager to get them, especially young Jewish Ashkenazi women, because they were in the population most likely to have the mutations. For many, the conflict about whether or not to find out what the future holds becomes unbearable: Do you take the test to learn whether you will probably develop cancer and discover that the same genetic shadow is likely to hang over your family and children? Do you decline to open this new window into your medical future? If you do open it, there is no way to close it again, and you have to live with the emotional and practical consequences for you and those you care about most. That includes the fact that this information will become part of your medical record, with unclear implications even for your employment prospects.

"Irma" was a vibrant young graduate student, full of hope for her future, happy in her studies, in love with her boyfriend, looking forward to a wonderful life, when she learned that she carried the BRCA1 mutation that she inherited from her mother. She had thought that it would be good to get the information so she had decided to be tested — but then found herself tortured by the results, miserable to be stuck with them, filled with regrets that she had opened the window into her double helix, wishing she did not know what she now could not put out of her mind. When she got the test results telling her that she carried the mutation she fell apart: she simply had not known ahead of time that she really did not want to know what her DNA contained, and that she would be devastated by what she learned.

Irma was not alone in her inability to predict how she would react to the test results. Can people be helped before they make their decision to better anticipate what they are likely to feel when they get the results of such genetic testing? That requires somehow pre-living the experience — and not with the unemotional, abstract, rational, and cool information processing we normally use for the distant future, but instead with the emotional processing we give to highly arousing experiences in the here and now.

Adequately informed consent is rarely obtained when people are wearing hospital gowns and ID bracelets, being prepped for the operating room early the next morning. Before surgery, somebody comes around with a clipboarded document full of small print that describes in much detail and medical jargon the multiple risks. This document makes it clear that just about everything could go wrong. You sign to agree that you won't blame the hospital, and that you are undertaking the procedure of your own volition, with fully informed consent. While there is little choice for medical interventions that are considered essential, it is a different situation for elective procedures like genetic testing.

In the early 1990s I consulted with psychologist Suzanne M. Miller at the Fox Chase Cancer Center in Philadelphia to develop a method that would improve informed consent for DNA testing. Suzanne and her colleagues were working with people at high risk for carrying the BRCA1 and BRCA2 mutations, and most of these people were eager and anxious for genetic testing to determine their risk for breast and ovarian cancer. Most, however, did not appreciate how the knowledge of their genetic predispositions might affect them. The usual genetic counseling practiced at that time involved an empathic but standard rational discussion of the alternatives, the options, the objective risks, and the uncertainties associated with each possible outcome and choice.

In the "pre-living" scenarios we developed for women considering genetic testing for their breast and ovarian cancer risk, we wanted each person to be able to anticipate her emotional reactions to the DNA revelations — not just abstractly, but by role-playing the experience as vividly, fully, and realistically as possible with the genetic counselor. We wanted to give these women an opportunity to anticipate and pre-live at least a mini-version of their likely hot reactions to the different test outcomes.

We proposed the following program. When a woman who is considering the test comes in for genetic counseling, she engages in realistic role-play with her counselor. The counselor tells her that the test results have come back from the lab, opens the folder on the desk, and reads that the results are positive: she has the mutation. In the safe, supportive environment with the counselor, she has the chance to express her feelings and thoughts, which might range from shock and disbelief to intense anxiety, despair, denial, anger, and questioning of the results. After these concerns are expressed and discussed, the counselor helps the woman begin to examine the options that are available and their likely consequences. Options include prophylactic mastectomy for the BRCA1 mutation, and prophylactic removal of the ovaries in the case of BRCA2. This candid pre-living experience goes on to deal with the long-term practical consequences affecting duration and quality of life, health care, insurance, employment and personal relationship prospects, childbearing, and any other concerns that arise.

This hot role-play experience is inevitably painful, but it gives the participant the emotional preview, as well as the cognitive information, to make a truly informed decision about whether or not to open the new genetic Pandora's box. The role-play also includes the happier scenario in which the genetic testing results are negative, and those implications are explored with equal depth and detail. After absorbing and reflecting on these pre-lived experiences, the decision to do or not to do the testing becomes the individual's well-informed choice.

As genome analysis and effective molecular science approaches to individually oriented diagnosis, prevention, and treatment continue to develop, decisions about looking into specific DNA prospects will become part of life for many. When these possibilities become realities, reaching wise decisions and giving informed consent about the range of preventive actions ideally should be guided by both the head and the heart, by the cool system and the hot system working together. The challenge will be to pre-live the emotions while also thinking coolly and proactively about what needs to be done.

WHAT DO YOU AND DON'T YOU WANT TO KNOW ABOUT YOUR FUTURE?

We differ enormously in how much we want to know about the risks and dangers we face. Imagine that while waiting in the doctor's office for a routine checkup, you are greeted by a medical researcher who wants you to answer a few questions. She asks you to vividly imagine scenes like this one: "You are on an airplane, 30 minutes from your destination, when the plane unexpectedly goes into a deep dive and then suddenly levels off. After a short time, the pilot announces that nothing is wrong, although the rest of the ride may be rough. You, however, are not convinced that all is well."

In the airplane situation, would you "listen carefully to the engines for unusual noises and watch the crew to see if their behavior is out of the ordinary?" Or would you "watch the end of the movie, even if [you] had seen it before?" The questionnaire doesn't disguise what the researcher is asking: do you want to know more about the stress you are facing or less? Another scenario runs as follows: "You are afraid of the dentist and have to get some dental work done." During the procedure would you want him to tell you what he is doing or prefer to do mental puzzles in your head? People who want to know more are categorized as "monitors"; those who would rather not know and prefer to self-distract or suppress are "blunters."

Women who were about to have a colposcopy, a common diagnostic procedure to check for the presence of abnormal (cancerous) cells in the uterus, were given the monitoring-blunting questionnaire while waiting to undergo the procedure, and they were then divided into two groups: monitors and blunters. In each group half were given extensive information about the procedure they faced, and half got the standard minimal information before they signed the typical consent form. The women reported what they felt before, during, and after the procedure, and the doctor as well as observers (who were "blind" to all the other information) assessed their psychophysiological reactions, including pulse, muscle tension, hand clenching, and expressions of pain and discomfort. Blunters who received minimal information and monitors who received extensive information experienced the least tension and stress during the exam and in the recovery period. Thus, when the amount of preparatory information they received fit their preferences, the women fared best and experienced the least stress.

These findings suggest that physicians should consider asking their patients how much they want to know about the medical options and decisions they face and the possible risks and benefits each choice entails. And in medical situations you might want to think about when you do or don't want to know more about the risks and side effects detailed in those consent forms or buried in tiny print in the warnings packed with each prescription drug. When do you want to monitor, and when do you want to blunt?

When facing stress, whether medical or social, monitors generally do better when they are told more, and blunters do better when they are told less. Matching the information to the individual's style reduces stress. As with all measures of individual differences, some people fall on either extreme of the spectrum, but most are more or less in the middle range. As a general rule for most people, if there is nothing you can do to reduce the stress because the situation is out of your control, monitoring typically increases anxiety and stress, and blunting tends to be more adaptive and self-protective.

LOOKING BACK AND AHEAD

It is a long journey from waiting for marshmallows in preschool to deciding how much current income goes into a retirement plan, overcoming cravings that threaten future health, and making well-informed medical decisions in the face of uncertain long-term consequences. But there is a common theme that connects the self-control challenges faced in all these diverse decisions over a lifetime. To resist a temptation we have to cool it, distance it from the self, and make it abstract. To take the future into account, we have to heat it, make it imminent and vivid. To plan for the future, it helps to pre-live it at least briefly, to imagine the alternative possible scenarios as if they were unfolding in the present. This allows us to anticipate the consequences of our choices, letting ourselves both feel hot and think cool. And then hope for the best.

11

PROTECTING THE HURT SELF: SELF-DISTANCING

Overcoming painful emotions like heartbreak and resisting temptations like cigarettes, unprotected sex, and unethical financial schemes all require cooling the hot system and activating the cool system. Both actions depend on the same two mechanisms: psychological distancing and cognitive reappraisal. This prescription is easy to write but tough to fill. The challenge is exemplified in the "Maria problem."

"Maria" was in a committed relationship with "Sam," and they had been together for nineteen years, since they were both graduate students. She was eager to have a baby almost from the very start, but Sam insisted "Not yet," and they kept postponing. One morning, with no forewarning, he announced that he had fallen in love with an undergrad at the university and he was leaving. Maria was heartbroken and struggled for months to get over the breakup, constantly replaying their last weekend together in her head. She could not grasp it. And she could not let it go.

Common beliefs held in Western cultures and most psychotherapies suggest that by facing her painful feelings honestly, Maria will gain insight and eventually be able to move on. In clinical practice, traditional psychotherapists usually urge their troubled clients to confront their unhappy experiences and feelings by persistently asking them, "I wonder why you felt that way?" But beginning early in the 1990s and continuing for twenty years, research by Susan Nolen-Hoeksema of Yale University revealed that while some people manage to get better by asking themselves the "Why?" question, many others get worse. They continue to brood and ruminate, only to become more depressed each time they recount the experience to themselves, friends, or empathic therapists. Instead of helping them "work through the experience," their endless rumination reactivates the emotional pain, reheats the anger, and reopens the wounds. In short, for many people, asking "Why?" doesn't help; it hurts.

When and why does this emotional confrontation backfire, and when does it succeed? That's the question that Ethan Kross could hardly wait to ask me in the fall of 2001, when, as a new graduate student, he walked into my lab at Columbia University. Answering that question is exactly what he has been doing ever since, starting with his studies at Columbia, which he completed in 2007, and after that in his research as a professor at the University of Michigan.

When Ethan and I first met, we brainstormed for many hours about how someone like Maria could be helped to cool her distress. We looked back at the marshmallow studies in which the preschoolers had pushed the treats and the bell as far away from themselves as possible, deliberately increasing the distance between themselves and the treat, which turned down their hot system and allowed their cool system to take over. Could this apply when adults try to overcome their anger and depression? It is easy to increase your distance from external stimuli like marshmallow temptations, but how do you create distance from your feelings and yourself?

LIKE A FLY ON THE WALL

As Ethan and I began to discuss different ways to help people self-distance when they are trying to overcome painful experiences, Ozlem Ayduk, who was then in the last phases of her graduate work in my lab at Columbia (and since then has become a professor at the University of California, Berkeley), became intrigued by the same question and joined us. Soon we conducted the first of many experiments on self-distancing. In this study, we enlisted Columbia University college students who had experienced a serious social rejection in an important close relationship that had caused them "overwhelming feelings of anger and hostility," and we asked them to reflect on it in one of two ways. Half of the students were invited to simply "visualize the experience through your own eyes . . . [and] try to understand your feelings." This was the "self-immersed" condition in which experiences are viewed as we normally see them through our own eyes. Most answers were emotionally hot, like these:

> *I was appalled that my boyfriend told me he couldn't connect with me because he thought I was going to hell. I cried and sat on the floor of my dorm hallway and tried to prove to him that my religion was the same as his.*

> *Adrenaline-infused. Pissed off. Betrayed. Angry. Victimized. Hurt. Shamed. Stepped on. Shitted on. Humiliated. Abandoned. Unappreciated. Pushed. Boundaries trampled upon. The worst communication.*

To create distance from the self, we asked the other half of the study participants to "visualize the experience from the perspective of a fly on the wall.... Try to understand your 'distant self's' feelings." From this "self-distanced" perspective, reactions were much less emotional, more abstract, and less egocentric:

> *I thought of the days and months running up to the conflict and was reminded of the academic stress and emotional turmoil I was going through combined with a lack of satisfaction with things in general. All these underlying currents and frustration led me to be irritable and thus sparked the conflict over a silly argument.*

> *I was able to see the argument more clearly.... I initially empathized better with myself but then I began to understand how my friend felt. It may have been irrational but I understand his motivation.*

The results were striking. When the participants analyzed their feelings from the usual self-immersed perspective, they recounted the concrete details as if they were reliving the experience (e.g., "He told me to back off" or "I remember watching her cheat on me") and reactivated the negative emotions they felt ("I was so angry, pissed off, betrayed"). In contrast, when they analyzed their feelings and the reasons for them from a distanced perspective, as a fly on the wall, they began to reappraise the event rather than just recounting it once again and reactivating their distress. They started to see it in a more thoughtful and less emotional way, allowing them to reconstrue and explain the painful past in ways that led to closure. Thus the same question — "Why did I feel that way?" — reactivates the hurt when one is self-immersed, but it will cool the hurt and provide a more adaptive narrative when one is self-distanced, like an observer. Before therapists ask their deeply self-immersed patients the "Why?" question, they might want to think about these results — and consider helping those patients reflect on experiences from a distance so that their hot system is not at its hottest, and their cool system can help them begin to think it through.

REAPPRAISAL FROM A DISTANCE

In a 2010 experiment, Ethan and Ozlem studied a new sample of participants and found that those who spontaneously distanced themselves when they reflected on their painful experience, and reappraised it rather than recounted it, felt better and became less stressed — not just in the short term, but also when they returned to the lab seven weeks later and were asked to reflect on the same experience again. To go beyond self-reports, another laboratory study done by Ethan and Ozlem showed that self-distancing helped reduce one of the most pernicious side effects of rumination: elevated blood pressure. When people think about painful negative experiences, particularly those that arouse intense feelings of anger and betrayal, their blood pressure rises. This becomes risky when blood pressure levels stay up over time. Ethan and Ozlem demonstrated that self-distancing effectively mitigates this harmful effect. The more people self-distanced, the more quickly their blood pressure returned to their typically healthy baseline levels.

Do the benefits of self-distancing when dealing with hurt feelings make a difference outside the relatively artificial conditions of laboratory experiments? Does self-distancing also help people solve problems and cope better with everyday conflicts in close interpersonal relationships? To address those questions, Ozlem and Ethan went on to do a large twenty-one-day daily diary study. At the end of each day of the study, participants logged in to a secure website that asked them to indicate whether they had had an argument with their partner that day. If they had, they were asked to reflect on their deepest thoughts and feelings about the event. Finally, they rated the extent to which they spontaneously self-distanced (i.e., adopted the fly-on-the-wall perspective) as they tried to understand their feelings surrounding their conflict with their partner.

Overall, people who spontaneously self-distanced when thinking about negative experiences in their relationship also used more constructive problem-solving strategies to resolve conflicts than those who did not spontaneously self-distance. Most interesting was that the low-self-distance people coped adaptively in conflicts, as long as their partners did not become negative and hostile toward them. But if their partners did become hostile, they fully reciprocated, sharply escalating the hostility. The combination of low-self-distancing people with highly negative partners became a formula for escalating hostility that was potentially toxic for the relationship's future. This pattern emerged whether conflict behavior was measured by self-report as it occurred during the diary study or by direct observation from independent raters when the partners discussed their conflicts in a laboratory setting.

Cognitive behavior therapists increasingly recognize that self-distancing is a prerequisite for therapeutic change for many people and many problems. They try to help their clients escape at least briefly from the self-immersed perspective by guiding them to realize that their beliefs and perceptions are constructions of "reality," not revelations of absolute truths that can be seen only one way. Clients learn how to step back from their feelings and actions and observe themselves from a distance. This is a prelude to exploring different ways of thinking about themselves and their experience that might prove to be more productive and less emotionally distressing. They learn that they can represent and think about events in alternative ways that can help them cool their distress. If you break a leg, for example, that's a fact you cannot change, as you discover when you try to walk on it. But you can change how you think about it: Is it a "horrible accident" that stresses you because you see all the things you can't do now, like jog and bike? Or is it an unexpected opportunity to do what you've long wanted to do, like catch up on reading the books you love?

James Gross at Stanford and Kevin Ochsner at Columbia have shown that similar reappraisal strategies can help people cool down a wide range of negative emotions. The researchers see these "cooling effects" not just in self-reports by participants indicating that they feel better when they use cooling strategies, but also in brain imaging studies. These studies show reduced activation of the hot system, particularly the amygdala, and increased activation in the prefrontal cortex when participants are reappraising intensely negative stimuli and experiences with the goal of cooling their emotional impact.

WHEN CHILDREN SELF-REFLECT

One of the joys of having many wonderful students and collaborators in research over the years is that when they get exciting results they connect with one another and the collaborations multiply. Angela Duckworth, a young professor at the University of Pennsylvania, was not my student, but our collaborations began when we met at a conference around 2002, each of us bringing our own students along. Subsequently, Ethan and Angela (as well as her student Eli Tsukayama, Ozlem, and myself) wanted to see if the effects of self-distancing found with adults would apply to children and young adolescents. It was a particularly important population to study, because it is the age when kids often torture one another with social exclusion and rejection, leaving those rejected feeling hurt, distressed, and angry. Too often the consequences turn into tragedies that result in public expressions of grief, but there's still little change in what children learn that could help them cope more constructively with the pain of rejection.

We focused especially on anger-related experiences and feelings in children because they had been linked in earlier research to destructive consequences, notably to escalating aggression, outbursts of violence, and the onset of depression. In the study by Ethan Kross and his team, fifth-grade boys and girls were cued to recall an interpersonal experience in which they felt overwhelming feelings of anger. They were instructed to "close your eyes. Go back to the time and place of the experience you just recalled and see the scene in your imagination." Then, in the self-immersed condition, they were asked to "replay the situation as it unfolds in your imagination through your own eyes." But in the self-distanced condition, they were instructed to "take a few steps back. Move away from the situation to a point where you can now watch the event unfold from a distance and see yourself in the event. As you do this, focus on what has now become the distant you. Now watch the situation unfold as if it were happening to the distant you all over again. Replay the event as it unfolds in your imagination as you observe your distant self."

Just as we had seen with young adults, self-distancing led the children to focus less on recounting and reliving the angry feelings that they had initially experienced and helped them rethink the event in ways that reduced their anger and promoted insight and closure. They developed a more objective perspective on the event, blamed the other person less, and created stories that helped them get over their anger. These findings came from a diverse sample of children, and they held regardless of gender, race, or socioeconomic status.

HEALING THE BROKEN HEART

Is the kind of pain Maria experienced from her "broken heart" just a metaphor, or does it capture a biological reality? That's another question about emotion regulation that Ethan Kross and his colleagues examined in a 2011 experiment. While their brains were being scanned by fMRI, people who had recently experienced an unwanted breakup viewed a photograph of their ex-partner and thought about their rejection. In another condition, the same individuals experienced intense physical pain from thermal stimulation to their forearm. During their physical pain, two brain areas (the secondary somatosensory cortex and the dorsal posterior insula) became activated; the same brain areas were activated when they thought about being rejected and looked at the image of the person who had broken their heart. When we speak about rejection experiences in terms of physical pain, it is not just a metaphor — the broken heart and emotional pain really do hurt in a physical way.

The overlap in how emotional pain and physical pain are experienced and processed in the brain raises many questions. One that is often asked, with tongue in cheek, is whether it would help to take painkillers to deal with heartbreaks and the endless other forms of rejection and exclusion. Researchers on social pain get this question at the end of their talks from people trying to be funny — but as it turns out, the answer is a strong yes! "Take two aspirins and call me in the morning" would be a coldhearted response to a friend's late-night report of fresh heartbreak, but it has a solid basis in the research.

Naomi Eisenberger and her colleagues at the University of California in Los Angeles gave volunteers either an over-the-counter (nonprescription) painkiller or a placebo to take every day over the course of three weeks. The volunteers monitored their levels of pain caused by social rejection in their everyday lives over those three weeks, unaware of whether they were taking the painkiller or the placebo. Those who were on the painkiller reported a significant reduction in their daily hurt feelings, beginning on average at day 9 and continuing to day 21, the last day of the study. Those taking the placebo showed no change. Another group of volunteers took either the painkiller or the placebo, again without knowing what they were taking, and then experienced a social rejection while in the fMRI scanner. While their brains were being scanned, they played Cyberball, a virtual reality game of catch from which they were eventually socially excluded: after seven tosses to them, they watched what looked like two other participants throw the ball to each other for forty-five throws without ever throwing it to them. In response to the social exclusion, those who had been on the painkiller for three weeks had significantly less neural activity in the pain areas of their brains.

If over-the-counter painkillers don't help soothe Maria's heartbreak, and she cannot manage the effortful mental acrobatics required to look at her experience as if she were the observant fly on the wall, another antidote remains. When feeling rejection pain, it helps to think about those to whom you are enduringly and securely attached. Just as looking at a picture of the person who rejected you can reactivate the pain of a broken heart, thinking about the people to whom you are deeply attached, people you love who love you back, can make it easier to overcome the kind of pain that kept Maria trapped in her past. This antidote is most effective for people who are already securely attached to others in their lives; it does not work as well for those who avoid attachment and close relationships.

12

COOLING PAINFUL EMOTIONS

THE MOST EXCITING FINDINGS from the marshmallow studies are not the unexpected long-term links between seconds of waiting on the Marshmallow Test and doing well later in life. More impressive is that if we have delay ability and use it, we are better protected from our personal vulnerabilities — such as a predisposition to gain unwanted weight, become angry, feel hurt and rejected, and so on — and can live with these predispositions more constructively. The research that shows how and why self-control has this positive effect has focused on a widespread and pernicious vulnerability called rejection sensitivity (RS), and I turn here to what has been learned about it.

THE FALLOUT FROM HIGH REJECTION SENSITIVITY

"High RS" people are extremely anxious about rejection in close relationships, anticipate abandonment, and often, through their own behavior, provoke the very rejection that they fear. If uncontrolled, the destructive effects of high RS can play out like a self-fulfilling prophecy. Imagine "Bill" as an exemplar of how severe RS can undo close relationships: he is both high in RS in his romantic relationships and low in delay and self-control ability. When his third marriage fails, he becomes depressed and anxious, and he tries talking to a therapist. Explaining the reasons for his last divorce, he angrily complains about his ex-wife's "lack of loyalty." The "evidence," in Bill's view, began with typical breakfast episodes. In his version, Bill was eager to talk and connect at breakfast every morning, but his wife was always still half asleep. Rather than listening attentively to him, she would yawn, close her eyes, and even turn away, gazing at the newspaper headlines or rearranging the flowers on the table. She was unresponsive to his complaints, he felt, and her uncaring behavior once "made me throw my damn scrambled eggs at her."

High RS people like Bill easily become obsessed with whether or not they are "really" loved, and their own ruminations further trigger a cascade of hot-system anger and resentment as their fears of abandonment escalate. In response to their distress, as well as the unhappy reactions of their partners, they become more coercive and controlling — openly or with passive aggression. They blame what they do on their partner's actions ("She made me throw my damn scrambled eggs at her") and they validate their fears of abandonment with the rejections that they at first imagine and then help create when their own rage erupts.

This signature pattern has predictable consequences, identified in research by Geraldine Downey and her students. Geraldine is a Columbia University psychology professor and has been my colleague since the early 1990s. She has long been the leader of research on the nature and consequences of RS. Her studies have shown that the relationships of high RS young men and women do not last as long as those who are low in RS. In middle school, high RS children are more easily victimized and bullied by their peers and are lonelier. In the long run, people who are high in this vulnerability continue to experience more rejection, which in time erodes their sense of personal worth and self-esteem, making depression more likely.

High RS not only undermines long-term relationships and inflicts hurt on others; it also biologically damages the people who have this sensitivity. Each time a person like Bill has outbursts and becomes enraged and stressed, his risk increases for cardiovascular disease, asthma, rheumatoid arthritis, a variety of cancers, and depression. Why?

Several experiments have assessed the immune system's physiological response to social rejection and have also examined brain activity during the response to rejection. When we feel rejected, neural activity and sensitivity increase in the dorsal anterior cingulate cortex and anterior insula. These regions are involved in the regulation of emotion, reward anticipation, and critical autonomic functions such as blood pressure and heart rate. In addition, when stress is experienced, the immune system produces inflammatory chemicals. In evolutionary history, there was good reason for the body to respond to stress by releasing inflammatory cytokines, the proteins that regulate the immune system in preparation for physical attack. That was and is adaptive because these proteins accelerate wound healing and therefore have great short-term value for recovery from physical injury. But when triggered long term, for example by chronic fear and anticipation of rejection, or by being unable to get over a huge rejection, the accelerated level of inflammation can lead to serious disease. Short-term inflammation in response to a wound was great for the survival of our ancestors, but long-term inflammation in response to hot system overreactions beginning at breakfast and lasting 24/7 is a prescription for illness.

HOW DELAY ABILITY PROTECTS

Soon after Geraldine came to Columbia, she and I, and our students, began a long series of joint studies to examine how self-control ability might protect high RS individuals against the unfortunate consequences of their vulnerability. The basic questions we asked were: Does delay ability protect against the negative effects of high RS? Do the same attention-control skills that let toddlers cope with the distress of even a brief maternal separation and help preschoolers wait for marshmallows also allow a high RS adult to cool himself before he becomes enraged when his wife pays attention to the newspaper headlines and not to him? RS was measured by how strongly participants felt that worries like these applied to them: "I often worry about being abandoned by others" and "I often worry that my partner really doesn't love me."

Ozlem Ayduk (who studied with Geraldine and me at Columbia at the time) led a study that looked at the preschoolers in the longitudinal studies I had begun at Stanford's Bing Nursery School. When they reached age 27 to 32, those high in RS who were not able to delay gratification as preschoolers on the Marshmallow Test had lower self-esteem, lower self-worth, and lower coping ability. They attained lower educational levels, used more cocaine/crack, and were more likely to get divorced. In contrast, the participants who were just as high in RS as young adults, but who had been able to delay gratification as preschoolers, were protected against these negative outcomes: their chronic anxiety about rejection did not become a self-fulfilling prophecy.

In 2008, a related study by the same team, again with Ozlem as the lead author, showed that people high in RS were also more vulnerable to developing features of borderline personality disorder. This disorder predisposes them to amplify minor disagreements and see them as personal attacks to which they react by becoming destructive to others as well as to themselves. And most important, those high in RS but also high in self-control ability were protected against these effects and preserved their relationships. We found this both in the follow-up of the preschoolers at Stanford and in two new samples, one of college students and the other of adults in the Berkeley, California, community. Overall, those who had high RS but also good self-control skills coped as well in their lives as those who were low in RS. When high RS people with good self-control skills were faced with stress and potential rejection in social relationships, they were able to use those skills to cool their hot, impulsive first reactions, thus restraining themselves from becoming enraged and aggressive and destroying their relationships.

The clearer and broader the connections became between what preschoolers did on the Marshmallow Test and what happened to them as their lives evolved, the more I kept asking myself: Would the results from Stanford, Columbia, and Berkeley hold up outside these privileged and selective communities? To find out, I needed a school as far as possible from the Stanford campus — both geographically and demographically.

FROM STANFORD TO THE SOUTH BRONX

It is hard to imagine a more extreme contrast than that between the sun-drenched, palm-tree-lined oasis of Stanford, California, in which the Bing preschoolers waited for marshmallows, and the South Bronx public middle school at which my students and I were eventually allowed to work. There was a tough defense system in place that closed New York's public schools to invasion and scrutiny by researchers, and it took four years of having every effort turned away before we got access to the school. This particular school's principal was willing to risk the wrath of the board of education and let us conduct research within his school's dark, fortresslike stone walls. This was in the early 1990s, when the city was barely beginning to recover from one of its worst slumps, and most of its public schools, including this one, were in deep decay. Classrooms were deteriorated, with plaster falling from the ceilings, high windows broken and barred, and dim lighting from bulbs that were as often burned out as they were lit. It was strikingly different not only from the Stanford-area public schools I had experienced through my own children, but from the public schools I had attended decades earlier in Brooklyn's working-class neighborhoods.

On my first visit, police cars were parked at the metal fences crowned with barbed wire. As the crowds of kids lined up and slowly filed through the metal detectors at the guarded entry doors, I was reminded of the times I had visited the high-security state prison during my doctoral training at Ohio State. Once inside the Bronx school, I was drawn by the roar of noise coming from the huge assembly hall, which was overflowing with loudly chattering and shouting students. Its aisles were patrolled by monitors, male teachers who marched up and down carrying batons and screaming louder than the students, over and over again: "Sit down and shut up!" On inquiry, I learned that this was the study period before classes began. This scene of sheer bedlam told me that we had found the school and sample we needed. As expected, it was a complete contrast to the Stanford school, but the conditions in the Bronx school were even more discouraging than I had anticipated.

We studied children as they entered middle school in the sixth grade at age 12 and followed them until they exited at the end of eighth grade at age 14. We did this in sequential waves over the five years of the project. When the sixth graders entered, they took the Marshmallow Test — except this time, the reward was lots of M&M's later rather than just a few right away. During the students' three years in the school, we collected various outcome measures so that we could see if what they did on the test did or did not predict their subsequent behavior.

Just as with the privileged Stanford children, the Bronx eighth graders who were high RS rated themselves as having lower self-worth and were rated by their peers and teachers as functioning more poorly. But again this correlation was found only for those young adolescents who had been unable two years earlier to effectively delay gratification on the Marshmallow Test. High RS did not condemn these children to interpersonal problems as long as they were able to cool their arousal and stress, as measured in the delay of gratification situation.

To follow how the children in the Bronx developed over time, we asked their peers to rate them on how well accepted they were socially and their teachers on how aggressive they were. The two sets of ratings turned out to correlate: kids perceived as more aggressive by their teachers were less accepted and rated more negatively by their peers. High RS youngsters were less accepted by their peers and seen as much more aggressive by their teachers, but only if they had been quick to ring the bell and settled for just a few M&M's.

The children who worried about rejection but were able to cool their stress and wait for their M&M's were perceived by their teachers as the least aggressive, and their peers saw them as the most socially accepted. The combination of high motivation to avoid rejection and self-control skills helped this group of children gain the acceptance they craved. High anxiety about rejection does not have to play out in a self-fulfilling prophecy. It can even help a rejection-sensitive child win the popularity contest.

I met "Rita" when she was 13 and in the seventh grade at the KIPP Academy Middle School in the South Bronx, the same KIPP school at which I met George Ramirez, who went on to Yale (Chapter 8). Rita spoke in a soft yet strong voice, and she reflected on everything she said. When she liked what she heard herself say, or thought it was funny, she broke into a big smile that lit up her face.

Rita had been at KIPP for three years, and before that she had been in the public school housed within the same structure. She had won the lottery needed to get into KIPP and her family met the poverty requirement. I asked Rita about her experience at KIPP, whose calm, serious, learning-oriented, well-disciplined classrooms were a different world from the chaos of the public school that shared the same building. She told me, "At first I didn't know how to adjust. Once I got here, I opened up. I started to talk to people. My teacher taught me I could write. So I have a notebook in which I write.... I like to write about my everyday life, not about how monkeys evolve."

Her face became serious. "I don't like to receive criticism. When I get criticism, I write it down. I write where it was, the person's name, what was said, why it hurt so much, and why it was said to me and not someone else. I show it to my counselor. She helps me get over it. I go talk to the person who criticized me and ask the questions I wrote down. It helps to talk and learn why it was said. It does relieve my anger. I learned everybody gets criticism. You just have to deal with it and move on."

Rita exemplifies the person highly sensitive to rejection but also able to self-control, who in outcome studies is functioning as well as those who are not especially rejection sensitive. With help, she is slowly managing to cool her rejection anxiety and get out of the self-immersed perspective, trying to gain distance from herself to externalize her hurt feelings by writing them down and discussing them. This allows her to learn how to get over these emotions and "move on."

When highly rejection-sensitive people feel angry and hostile, as they often do, they have an advantage if they also have the ability to cool and slow themselves down by taking a deep breath, regulating their thoughts strategically, and thinking of their long-term goals. They can make these strategies automatic rather than effortful if they develop and practice *If-Then* implementation plans that connect their hot triggers (*If* she reads the newspaper) and internal cues (*If* I am starting to feel angry) to their self-control strategies (*then* I take a deep breath and start counting backwards from 100).

Those delay skills can also be used to cool the aggressive

impulse by activating a hot thought that is incompatible with it. For example, if someone like Bill developed better self-control skills, he might be able to vividly imagine how his egg throwing in the heat of the moment would have him reading a "Dear Bill" letter when he returns home that night and finds his wife's closet empty. The mechanisms at work here, when delay ability allows the split-second pause of reflection before action, are the same as those that help people with other vulnerabilities (tendencies toward borderline personality disorder, obesity, or drug addiction) better regulate and control their behavior.

In the *Journal of Pediatrics* in 2013, Tanya Schlam and her colleagues reported that the amount of time the Bing Nursery School preschoolers at Stanford waited on the Marshmallow Test predicted their body mass index thirty years later: "Each additional minute that a preschooler delayed gratification predicted a 0.2-point reduction in BMI in adulthood." The authors rightly caution that a significant correlation, while impressive and rare over such a long time period, does not imply a causal link. It can help, however, encourage researchers, educators, and parents to continue to develop interventions to improve self-control skills in young children.

SELF-CONTROL IN DUNEDIN

A scientist is always eager for independent replication of research findings, preferably in different populations and contexts. In 2011, I was reassured to learn that parallel results about the protective effects of self-control early in life were being found by another research team that worked with a very different population on the other side of the globe, decades after the marshmallow studies began. Terrie Moffitt, Avshalom Caspi, and their colleagues looked closely at the lives of more than a thousand children born in Dunedin, New Zealand, in a single year and followed them over the years to see how they were doing at age 32. They used measures of self-control and long-term outcomes different from ours. They assessed self-control during the first decade of life with a wide variety of observational ratings, as well as with reports from parents, teachers, and the children themselves. They asked about aggression, hyperactivity, lack of persistence, inattention, and impulsivity. To assess health outcomes, they measured substance dependence, smoking, and metabolic abnormalities (such as obesity, hypertension, and high cholesterol). They looked at wealth outcomes, including income levels, family structure (such as single-parent child rearing), saving habits, credit problems, and financial dependence. They assessed antisocial behavior, such as criminal convictions. Regardless of the measure used, poor childhood self-control significantly predicted negative adult outcomes: worse health, more financial troubles, and more crimes committed.

It was good to see how consistent the 2011 Dunedin findings in New Zealand were with those that began in the Surprise Room at Stanford in the late 1960s: self-control, especially early in life, has predictive value. More important, as the other research in this chapter has illustrated, it has protective value, helping prevent dispositional vulnerabilities from playing out destructively. That makes self-control skills worth nourishing in our children, and in ourselves.

13

THE PSYCHOLOGICAL IMMUNE SYSTEM

WHEN OUR SELF-CONTROL EFFORTS fail, we have a hidden ally that in time helps us feel better, or at least not too awful, no matter how badly we mess up or how harshly life treats us. Evolution provides us with automatic protective mechanisms that come to the rescue when life deals us terrible blows we simply cannot control, and when our strengths are insufficient, our cool system too tired, and our own fallible behaviors and fragile feelings get us in trouble.

These mechanisms used to be called ego defenses, but early in this century Daniel Gilbert at Harvard, working with Timothy Wilson from the University of Virginia and others, has broadened, revised, and more aptly renamed them the "psychological immune system." This system creates a safety net to protect us from the effects of chronic stress, and it fortifies us so that we can cope with terrible news — like a routine checkup that turns into a cancer diagnosis, a plunge in retirement funds, a pink slip announcing that it is time to clear out the office, or the sudden death of a person we love. While the biological immune system keeps us alive by protecting us against illness, the psychological immune system reduces perceived stress and helps us avoid depression. The stress-reducing and antidepressive effects of the psychological immune system bolster the biological immune system, and the two continuously interact to try to keep us smiling and healthy even when life is especially harsh.

PROTECTING SELF-REGARD: SELF-ENHANCEMENT

The psychological immune system finds ways for us to avoid hating ourselves for bad outcomes and credit ourselves for the good ones. It lets us attribute the bad outcomes to everything from the government, an incompetent underling, or a jealous colleague to a moment of bad luck or some other condition outside our control. It helps you fall asleep at night after reliving a work episode in which a colleague referred to your idea at the group meeting as a formula for disaster. OK, you think, perhaps it wasn't such a good idea, but it's forgivable because you were coming down with the flu. As social psychologist Elliot Aronson put it in the title of his book with Carol Tavris, *Mistakes Were Made (but Not by Me)*.

The psychological immune system preserves our sense of being good, smart, and worthy. Provided we are not severely depressed or dysfunctional, it allows us to see ourselves as having more positive and fewer negative qualities than most of our peers. It does not work this way with everything, though: you may see yourself as intelligent overall but incompetent with technology, or as being good at self-control when it comes to work but not when it comes to chocolate. Nevertheless, when people rate themselves on Shelley Taylor's "How I See Myself" questionnaire, which lists 21 qualities including "cheerful," "academically able," "intellectually self-confident," "sensitive to others," and "desire to achieve," between 67 and 96 percent rate themselves better than they rate their peers. David G. Myers, a social psychologist at Hope College, captured the gist of the multitude of studies on self-evaluation.

> In one College Board survey of 829,000 high school seniors, zero percent rated themselves below average in "ability to get along with others," 60 percent rated themselves in the top 10 percent, and 25 percent rated themselves in the top 1 percent. Compared to our average peer, most of us fancy ourselves as more intelligent, better looking, less prejudiced, more ethical, healthier, and likely to live longer — a phenomenon recognized in Freud's joke about the man who told his wife, "If one of us should die, I shall move to Paris."...
>
> In everyday life, more than nine in ten drivers are above average drivers, or so they presume. In surveys of college faculty, 90 percent or more have rated themselves as superior to their average colleague.... When husbands and wives estimate what percent of the housework they contribute, or when work team members estimate their contributions, their self-estimates routinely sum to more than 100 percent.

We cannot all be above average. The important question is whether this illusion of self-regard is ultimately good or bad for us. Should we cheer this type of self-enhancement, give it a positive name like "self-affirmation," be glad to see it in our children, and not censor it in ourselves? Or is this overevaluation of the self a neurotic mechanism, a defense system that we need to overcome so that we can see ourselves more accurately? Not surprisingly, consistent with the phenomenon itself, advocates on each side are passionate about the perceptiveness of their own view and the foolishness of the opposition's. Shelley Taylor and her colleagues explored the impact of self-regard in a series of experiments beginning in the late 1990s and continuing for many years, and their results brought new evidence into the debate.

Taylor and her team demonstrated that high self-enhancers, the people who get higher self-affirming scores when they compare themselves with peers, in fact have lower chronic biological stress levels. Biologically, this happens in large part through the work of the hypothalamic-pituitary-adrenal axis (HPA), which regulates everything from digestion and temperature to mood, sexuality, physical energy, and the biological immune system. The HPA also indicates how well or how poorly you react to stress and trauma. High self-enhancers have a healthier HPA axis profile than low self-enhancers. They are better able to attenuate the hot system when reacting to threats because their calming parasympathetic activity increases, as does their comfort level. This reduces stress, putting high self-enhancers in a more self-soothing, recuperative mode, in which they can restore themselves and heal rather than having to tense up for the next battle — whether to face the wild hyenas of our ancestors' time or their contemporary versions today.

These findings contradict the traditional belief, still shared by many psychotherapists, that positive illusions and self-enhancement are defensive denials of negative personal characteristics and signs of grandiosity and neurotic narcissism, and that the effort to suppress or repress one's negative qualities has large biological costs. In fact, positive self-affirming mental states, including positive illusions (as long as they are not extreme distortions of reality), enhance healthier physiological and neuroendocrine functioning and lead to lower stress levels. The realists who perceive themselves more accurately experience lower self-esteem and more depression, and they are generally less mentally and physically healthy. In contrast, healthier individuals perceive themselves with a warm, even if somewhat illusory, glow.

There are close parallels between the workings of the psychological and the biological immune systems. Both serve us well, but both can backfire if they either overreact or underperform. Each has to strike a balance between two competing needs, as Daniel Gilbert points out. The biological immune system has to identify and kill foreign invaders like viruses, but it has to avoid killing the body's good cells. Likewise, it may be adaptive and good for self-esteem if the psychological immune system leads you to think you are better than most peers, but it's a different story if you believe you are better than everybody else.

Even if the psychological immune system is doing a fine job balancing self-enhancement and realism, it often makes us incorrectly predict how we would feel if terrible things were to happen. If we are asked to imagine how we would feel if we became paraplegic, we are apt to anticipate a terribly unhappy life, as Gilbert and other researchers have shown. If it actually happens to us, our psychological immune system fortunately helps us make the best of it, and we soon wind up feeling much better than we thought we would. The downside of this system is that it makes us poor predictors of our future happiness; the upside is that it makes us better survivors when life goes badly. But what happens when the psychological immune system fails us?

LOSING THE ROSE-COLORED GLASSES

Aaron Beck, a pioneer in the development of cognitive behavior therapy beginning in the 1970s and continuing into this century, proposed that the severely depressed suffer from an unrealistically negative view of the world, of the self, and of the future. He conceptualized depression as a generalized negative mental set, like a pair of dark glasses that turns everything into gloom. But might a negative self-image partly reflect depressed people's realistic recognition of their own lack of positive interpersonal skills and competence? Perhaps the depressed actually *are* less socially skillful and therefore are perceived more negatively both by other people who observe them and by themselves.

To untangle these possibilities, I worked with Peter Lewinsohn and his colleagues from the Psychology Clinic at the University of Oregon in 1980 to examine how clinically depressed patients evaluate their performance. We needed to get both the self-ratings of depressives for their actual performance in social interactions and the ratings of independent observers who watched their performance, so that we could assess their congruence. Then we compared these patterns in the depressed patients with those in psychiatric patients who had equally severe mental problems but were not depressed, and also with those in non-patient control participants who had no current or past depression problems (but who were similar in age and demographics).

Participants were seated in small groups in a comfortable informal seating arrangement and were told that the researchers wanted to learn more about how strangers related to one another. Each person in these small group meetings introduced himself or herself with a short monologue, and they were left alone to converse for 20 minutes. The observers, carefully trained and blind to the diagnoses and histories of the participants, rated what they observed from behind one-way mirrors on standard rating scales that listed many desirable attributes: friendly, popular, assertive, attractive, warm, communicates clearly, socially skillful, interested in other people, understands what others say, humorous, speaks fluently, open and self-disclosing, has a positive outlook on life, and so on. Right after each session, the participants rated their own performance in the group interaction on the same scales used by the observers.

The depressives, far from seeing themselves through dark lenses as we had presumed, were cursed by twenty-twenty vision: compared with other groups, their self-ratings of positive qualities most closely matched how the observers rated them. In contrast, both the nondepressed psychiatric patients and the control group had inflated self-ratings, seeing themselves more positively than the observers saw them. The depressive patients simply did not see themselves through the rose-colored glasses that the others used when evaluating themselves.

During the next few months, while they were treated with cognitive behavior therapy at the University of Oregon's Psychology Clinic, the depressive patients began to enhance their self-evaluations, gradually rating themselves as more socially competent. Although the observers did not know that treatment was taking place, they also began to rate the depressives more positively. But even though the depressives saw themselves more positively after treatment, they were still more realistic in their self-evaluations and saw themselves more like others saw them. Importantly, the differences in the self-ratings between the three groups declined: the depressives were feeling better, and presumably their bolstered psychological immune systems raised their levels of self-evaluations.

If the observers — who were the criterion for accuracy in this research — had been asked to rate themselves, they probably also would have tended toward inflation, just as the participants in the control group did. We see others accurately, but we wear the rose-colored glasses when we rate ourselves, if we are fortunate enough to not be depressed. In fact, this kind of inflation in self-evaluation may be what helps protect most people from being depressed.

HOW FEELINGS TWIST THINKING

What astonishes me, no matter how often I see it, is the power with which strong negative emotions can trump cool thinking. They can create fallout that distorts not just what we experience in the moment, but also what we expect in the future and how we evaluate ourselves. To examine how this plays out, Jack Wright and I studied how happy and sad feelings impacted performance on a challenging problem-solving task. Jack, who had been my student at Stanford and is now a professor at Brown University, asked college student volunteers in one condition to imagine, in vivid detail, a situation that would make them feel very happy, while those in another condition imagined a situation that would make them very sad. They were encouraged to picture the surrounding people and objects in their "mind's eye," to see the sights, hear the sounds, experience the event, think the thoughts, and have the feelings they would have had if they were really there. For example, to induce a happy mood, one student imagined future graduation from law school and fantasized about himself on graduation day, "long awaited and strived for, standing there knowing that I did it, I finally did it." To create a sad mood, another student imagined "I was rejected at every law school I applied to."

While maintaining their mood states, participants had to match pairs of rotating three-dimensional figures shown on the computer in various angles of rotation, ranging in difficulty from very easy to unsolvable. Over many trials they received false but completely credible feedback indicating that they were either highly successful or failing on the most difficult problems. The most striking finding was the unfortunate effect of the combination of feeling sad and believing that they were failing. The students in a sad mood greatly overreacted to their negative performance feedback, lowering how they evaluated their own performance and their expectations for the next set of tasks much more sharply than those who got the same feedback but who were in a positive mood. Students who had been induced into a happy mood formed much higher expectations for their future performance, recalled more of their successful experiences, and made more favorable self-descriptions. They evaluated themselves as more intelligent, attractive, self-confident, popular, successful, and socially skilled, and they had higher expectations about their future performance than those who had self-induced negative emotions.

DINNER WITH JAKE

I try to keep the evidence of the benefits of self-enhancement in mind when I think about "Jake." I was once stuck at a formal dinner next to Jake, a self-made man who had amassed a fortune in the financial world. His self-enhancement was so excessive that while it may have made him immensely successful by many counts, it also made him unbearable, at least to my hot system. Convinced he was fascinating, he told me stories nonstop about his special qualities, beginning with the pheromones released in his natural bodily sweat that, according to him, made attractive young women eager to be with him.

Given the demonstrated benefits of self-enhancing, I kept wondering why I so quickly disliked Jake, who was in my eyes the prototype of extreme self-affirmation. Perhaps high self-affirmers are healthier but friendless. Might the self-enhancers turn off other people by being too self-absorbed and having too little empathy? Might they be too busy enhancing themselves to perceive what is going on in the minds of the people in front of them? When researchers asked those questions, they found that people who view themselves more favorably than their friends view them had friendships that were just as long lasting, strong, and positive as those of low self-enhancers.

Then what went wrong at that dinner? Most adaptive self-enhancers make subtle and automatic discriminations about the situations in which public self-enhancement is and is not appropriate, and where modesty is or is not expected. We usually self-enhance in our own heads, nourishing self-regard and self-soothing privately, not publicly. From the thin slice of Jake's behavior that I endured, his problem seemed to be that he was indiscreet in when and where he self-enhanced. I suspect that his indiscretion was related to another deficit: a poorly developed theory of mind (ToM).

As previously discussed, ToM is an important mental ability that begins in early childhood and allows us to understand that our beliefs may be false, that the way things appear may not mirror reality, and that other people may not perceive the same scene or event the way we do. In normal development, preschoolers already exhibit ToM, and it is strongly related to their ability to suppress impulsive responses. If Jake's goal was to impress me, then his ToM was not working well; but maybe his goal was to impress himself, and his ToM could not care less. Unlike Jake, people whose self-enhancement is coupled with the desire to also make other people feel good about themselves have a great advantage: they can build mutually supportive and satisfying close relationships that not only have their own obvious benefits but also enhance their individual strengths and self-regard.

ASSESSING THE PSYCHOLOGICAL IMMUNE SYSTEM

The psychological immune system that promotes high self-regard and links to good mental and physical health was seen as a brittle neurotic defense system by many psychotherapists from the time of Freud into the 1990s. Therapists often tried to help people dismantle this system and get over their defenses. And this is still the case today with some therapists: if you enter a psychotherapist's office now, without knowing that professional's background, orientation, and training, there is a good chance that your self-enhancing system will be treated as a problem to be overcome rather than a strength to be embraced. But therapists trained in cognitive behavior therapy — the current evidence-based approach to treating psychological problems — are likely to take the opposite approach. Typically, they will work to strengthen the psychological immune system, while also helping to control its excesses.

While health psychologists, cognitive neuroscientists, and behavioral researchers have demonstrated the value of the psychological immune system and the personal qualities that keep it healthy, behavioral economists and many psychologists have shown its downside. They find that, unless kept very carefully in check, optimism, self-affirmation, and the related positive qualities generate a bias that leads to overconfidence and potentially dangerous decision making and risk taking — across virtually every profession and business examined closely. No matter how careful the screening, and how impressive the individual's track record, the optimistic bias of "Yes, I can!" (it also comes in the form of "Yes, I know!") leads these highly skilled, successful professionals to take on excessive risks — even when they are honest, well-trained, well-intentioned models of lifelong rigorous self-control and self-discipline. These risks can easily end in disaster, and the people who are vulnerable to making such mistakes periodically bring their own success to a screeching halt when their overconfidence leads them to breach social norms and ethics, which often lands them in the headlines.

The scandal of General Petraeus, director of the U.S. Central Intelligence Agency, illustrates the power of perceived immunity from consequences. This is an instance of the hot system going full force — even though the probability of exposure is glaringly obvious to the cool system. Four-star general David Petraeus was widely esteemed, serving as the poster boy for cool cognitive control. He personified Spartan self-discipline, including daily jogs at dawn for many miles in the hills of Kabul when he commanded the troops in Afghanistan. He was appointed head of the CIA in September 2011, but was quickly brought down in November 2012 by a long string of emails that revealed details of his adulterous love affair with his biographer. The correspondence was uncovered by the FBI, and it led to the general's immediate resignation. The tragic irony of his situation (or, depending on your perspective, its absurdity) is Shakespearian.

HUBRIS: THE ACHILLES' HEEL

The Petraeus story reminds us that even the almost invincible hero Achilles, of Greek mythology, had a vulnerable heel, the one exposed hot spot that could bring about his downfall, and that made him human. Nevertheless, while recognizing that we all have hot spots that make us vulnerable, we still expect that people who are excellent at self-control will also be more alert and sensitive to delayed long-term risks.

As discussed, high delayers are better protected against experiencing stress, and this in turn can make them less sensitive to danger signals. Likewise, because they tend to experience more success and mastery over their life course — from better physical health to higher financial gains — they may be more predisposed to some costly decision biases, particularly as a result of the illusion of control. As Petraeus's story illustrates, the illusion of control can cause a formidably competent, high self-control person to reveal information over email that can undo the successful life he built.

The consequences of the illusion of control can be catastrophic, particularly in some financial risk-taking situations, when high self-controllers may feel in control and then fail to react appropriately to external feedback and danger signals. This happened in the real world during the financial disaster of 2008. In 2013, it was simulated and analyzed by Maria Konnikova at Columbia University in five experiments on risk taking when money was at stake, albeit not in the billions. Staying calm, optimistic, and self-confident, the high self-control decision makers disregarded the feedback about their losses, were shielded from stress, and lost more money than the low self-controllers, who became anxious sooner, responded to the feedback, and quit before they went broke. In the end, in some conditions, it is the low self-controllers, with their lessened confidence and heightened anxiety, who can end up ahead.

The benefits, however, may not last. The researchers induced heightened illusory control in the low self-control participants by having them succeed in predicting coin flips, or getting them to recall times when they had made good decisions and had been in high-control situations. Feeling more confident, these participants quickly lost their initial advantage: they started to resemble the high self-controllers — and to make the exact same poor choices (and lose money) as a result.

FROM BEDROOM TO BOARDROOM TO BURNED FEET

Reviewing the paradoxical literature in which the "I think I can!" optimists not infrequently mess up their lives and the lives of people who depend on them, Daniel Kahneman, a Nobel laureate in economics and my colleague in psychology, points out, "An optimistic bias plays a role — sometimes the dominant role — whenever individuals or institutions voluntarily take on significant risks. More often than not, risk takers underestimate the odds they face, and do not invest sufficient effort to find out what the odds are." He then presents powerful evidence that optimism creates enthusiastic inventors and energetic, hard-working, courageous entrepreneurs who are eager to seize the day — but whose confidence also fosters their delusions and leads them to minimize the risks and suffer costly consequences. Asked about the probability of success for "any business like yours," one-third of American entrepreneurs said their chance of failing was zero. In fact, only about 35 percent of such businesses in the United States survive for five years. This seems to hold for everything from a small bed-and-breakfast venture to a Silicon Valley start-up promising the next big thing. At the very least, it may be reassuring that optimistic entrepreneurs are even more likely to take excessive risks and make unsound bets with their own money than with other people's money.

Thomas Astebro, a researcher who studied the fates of almost 1,100 new inventions submitted by eager innovators, found that less than 10 percent of them reached the market, and of those that did, 60 percent got negative returns. Half of the inventors withdrew after receiving objective reviews predicting that their inventions were sure to fail, but 47 percent of the remaining half persisted, doubling their initial losses before quitting.

Six out of the roughly 1,100 inventions scored big, however: they yielded returns exceeding 1,400 percent. Those are the kinds of high but extremely unlikely and unpredictable payoffs that cause unrelenting optimists to continue buying lottery tickets. Those odds also keep them pulling the levers on the slot machines and rolling the dice after performing little rituals to increase their luck in the gambling casinos. A schedule of reinforcement that delivers a big payoff at rare, unpredictable times can, in experiments, keep pigeons continuously pecking on a lever forever, as B. F. Skinner and his students demonstrated, and it can seduce gamblers to keep losing until they can't get another loan. It also gets optimistic entrepreneurs and innovators to keep working thousands of hours in the hope that they will become the next billionaire.

The dangers and costs of overconfidence are not restricted to the world of entrepreneurship and financial risk taking. They apply equally to any experts who are optimistic enough to make predictions about outcomes that are subject to chance or largely unknowable. In one study, for example, the diagnoses that highly competent physicians made while their patients were still alive in the hospital's intensive care unit were compared with what their autopsies later revealed. Physicians who had been "completely certain" of their diagnosis turned out to be wrong 40 percent of the time.

Early in my career, I lost many friends in clinical psychology by calling the field's attention to the discrepancy between the confidence with which clinicians predicted outcomes, like the probability that particular psychiatric patients would return to the hospital within a few years, and their consistently stunning lack of validity. The predictions of renowned expert diagnosticians were no more accurate than those made by untrained bystanders. The weight of the patients' folders summarizing their past psychiatric history was by far the best predictor of the incidence and speed of their rehospitalization, greatly beating any combination of the best tests, extensive interviews, and expert clinical judgments.

I discovered the problem of unwarranted confidence in experts' predictions not just by reviewing other people's failures. I found it in my own research. I worked with the first Peace Corps project that sent young volunteers to teach in Nigeria in the early 1960s. While the volunteers were undergoing training at Harvard, we used a costly and elaborate assessment process that relied heavily on interviews with trained experts, faculty ratings, and a battery of cutting-edge personality tests. In a final meeting that lasted many hours, an assessment board representing experts from a variety of disciplines and with diverse experiences met to discuss each individual volunteer and reach a consensus about his or her personality characteristics and likely success in a teaching position. There was high agreement among the ratings from these diverse sources, and the assessors were confident about their usefulness for predicting how well each of these volunteers would do on their assignments in the field.

A year later, the assessment board's predictions proved to have zero validity: they were not significantly correlated with the performance ratings reported by the volunteers' supervisors in Nigeria. In contrast, the candidates' simple self-reports about their attitudes, qualities, and beliefs had at least moderate predictive value. While this experience was shocking at the time, in retrospect it turned out to be prophetic: a similar lack of validity for expert predictions made in this manner has been shown to be the rule, whether in long-term forecasting for the stock market, the behavior of psychiatric patients, the success of business enterprises, or virtually any other temporally distant outcome, as fully documented in Kahneman's 2011 book, *Thinking, Fast and Slow.*

In short, the psychological immune system protects us from feeling too bad when our predictions fail, but it can also keep us clinging to beliefs in the face of evidence that persistently contradicts them, leading us to make mistakes that have high costs. Optimistic illusions can be hard to disconfirm, even when they burn your feet. In July 2012, in San Jose, California, twenty-one people had to be treated for burns after attempting to walk on hot coals, as inspired by a motivational speaker extolling the power of positive thinking. Burned feet notwithstanding, in further testimony to the power of the psychological immune system and the human ability to reduce psychological dissonance, many of the people who had tried it apparently felt, after they cooled their feet, that it was a transformative positive experience. Even when the prefrontal cortex doesn't protect us, and "I think I can!" leads to burned feet, the psychological immune system keeps doing its work.

14

WHEN SMART PEOPLE ACT STUPID

AS IMPEACHMENT LOOMED FOR the president of the United States in 1998, a reporter called me to ask if we could trust what President Clinton did when he was working at his desk in the Oval Office, now that we knew what he did under it. Other reporters were less direct, but they had the same concern. Their questions reflected the common belief that qualities like self-control, conscientiousness, and trustworthiness are broad traits that characterize a person's behavior not only stably over time but also consistently in many different kinds of situations: it assumes that a person who lies and cheats in one kind of situation is also likely to be dishonest in many other situations, whereas one who is conscientious will be predictably conscientious in diverse contexts. These expectations are violated each time the headlines announce the fall of another famous person in a position of public trust who turns out to have a secret life, exposing a side of his personality that appears to be the opposite of his public self. Predictably, a torrent of speculations follows that always raise the question: "Who is he *really?*"

President Clinton's pattern was hardly unique. One of the most stunning examples of such an inconsistency in behavior was the fall of Sol Wachtler from his position as chief judge of the State of New York and the New York Court of Appeals to incarceration as a felon in federal prison. Judge Wachtler was famously revered for advocating laws to make marital rape a punishable crime, and was deeply respected for his landmark decisions on free speech, civil rights, and the right to die. After his mistress left him, however, the judge reportedly spent months harassing her, writing obscene letters, making lewd phone calls, and threatening to kidnap her daughter. How did this model of jurisprudence and moral wisdom turn into a handcuffed prisoner on his way to jail? Judge Wachtler attributed his own behavior to his problems with an uncontrollable romantic obsession. One expert asked about Wachtler suggested that he might have a brain tumor the size of a baseball. He didn't.

The headlines announce similar stories about celebrities and public figures in the entertainment world, religious institutions and pulpits, business, sports, and academia — no area is exempt. Tiger Woods, the golf star hero, was the personification and ideal of rigorous self-discipline, not just in mastering his physical skills but in his sensational capacity to focus his attention. He was a supposedly happily married man, but he ultimately confessed to a private life with mistresses that violated his well-cultivated public image. The sports idol suffered one of the more memorable instant falls from grace, or at least from public celebration — for a while. In time, his descent was followed by that of the world champion marathon cyclist Lance Armstrong, whose career and extraordinary life were tainted by a doping scandal.

CONTEXTUALIZED SELF-CONTROL

"How do you explain these folks?" reporters invariably ask.

They want a short answer for their deadline. I give them the shortest version: President Clinton had the self-control and delay ability to win a Rhodes scholarship, attain a Yale law degree, and be elected to the U.S. presidency, apparently combined with little desire — perhaps no ability, and certainly no willingness — to exert self-control for particular temptations like junk food and attractive White House interns. Likewise, the judge and the golf star had the self-control skills to excel in the pursuit of their most important career goals, but not in other contexts. To be able to delay gratification and exert self-control is an *ability*, a set of cognitive skills, that, like any ability, can be used or not used depending primarily on the motivation to use it. Delay ability can help preschoolers resist one marshmallow now to get two later, but they have to want to do that.

Whether or not self-control skills are used depends on a host of considerations, but how we perceive the situation and the probable consequences, our motivation and goals, and the intensity of the temptation, are especially important. This may seem obvious, but I emphasize it here because it is easily misunderstood. Willpower has been mischaracterized as something other than a "skill" because it is not always exercised consistently over time. But like all skills, self-control skill is exercised only when we are motivated to use it. The skill is stable, but if the motivation changes, so does the behavior.

Many celebrities and public figures exposed in the headlines probably did not want to resist their temptations. On the contrary, they often seemed to expend considerable effort seeking and pursuing them. Their optimistic illusions and inflated self-worth, shared with the rest of humanity but perhaps even more grandiose in them, made them feel invulnerable. They did not expect to be caught even if they had been in the past. They also believed that if they were discovered, they could still get away with it — which is not an unreasonable expectation for some, given their past experiences. Their histories of success and power may also encourage them to spin entitlement theories that exempt them from the usual rules and encourage them to do what less powerful people can't. As Leona Helmsley, New York's billionaire ex–hotel queen, was alleged to have said before beginning her prison term, "Only the little people pay taxes." If they don't make remarks like that, their prospects for redemption usually remain excellent even if they are exposed. Modern-day fallen heroes often rise, phoenix-like, from the newspaper ashes that announced their downfall to host television shows, run news and interview programs, or become well-compensated consultants.

The ability to exert self-control and wait for marshmallows implies neither that it will be exercised in every domain and context nor that it will be used for virtuous goals. People can have excellent self-control skills that they use creatively for good purposes valued by society. They can also use the same skills to create hidden families, offshore bank accounts, and secret lives. They can be responsible, conscientious, and trustworthy in some areas of their lives and not in others. If we look closely at what people really *do*, not at what they say, across different situations with regard to any dimension of social behavior, it turns out that they are *not* very consistent.

THE CONSISTENCY PARADOX

When we look at the people we know, nothing is more obvious than that they differ greatly in their social behavior and characteristics, on whatever dimension we consider. In general, some are much more conscientious, sociable, friendly, aggressive, quarrelsome, extraverted, or neurotic than others. We make these judgments easily, and we mostly agree not only with one another but also with the self-perceptions of the people we are judging. These widely shared impressions of what we are like are very useful, indeed vital, for navigating the social world, and they allow us to make reasonable predictions about what to expect from other people.

Situations also influence social behavior in an important way, depending on how they are perceived. Regardless of how conscientious a person tends to be or not be, most will be more conscientious about being on time when picking up the kids from preschool than when meeting a friend for coffee, and they will be more sociable and extraverted at large parties than at funerals. That kind of variability is evident.

The conception of human traits, however, makes an additional assumption — namely that an individual will be consistent in the expression of a trait across many different kinds of situations in which the trait is desirable. It is assumed that a highly conscientious person will be more conscientious than a less conscientious person consistently across many different kinds of situations. If Johnny is judged to be much more conscientious than Danny "on the whole," then he should also be ranked higher than Danny on his completion of homework assignments and on his attendance record, as well as on how conscientiously he keeps his room organized at home and how trustworthy he is when he babysits his younger sister. Is this assumption justified? Does the person high in any important psychological trait generally remain above the person low in that trait across many different kinds of situations?

The assumption that people are broadly consistent in what they do, think, and feel in very different situations is intuitively powerful. It is fed by the hot system, which quickly forms impressions from the smallest slices of behavior and generalizes them to anything that can more or less fit. But does it hold up when we use the benefit of the prefrontal cortex to look closely at what people really do across different situations — whether it is President Clinton, our family and friends, or ourselves?

When I was preparing to teach my first course on assessment as a new lecturer at Harvard, I started asking these questions: Can you use your work partner's conscientiousness at the office to predict how conscientious she will be at home? Can I predict how my colleague — who is known as "the loose cannon" at department meetings — behaves at home with his children? To my own surprise, study after rigorous study failed to support the core trait assumption: people high in a trait in one kind of situation often were low in that trait in another type of situation. The aggressive child at home may be less aggressive than most when in school; the woman exceptionally hostile when rejected in love may be unusually tolerant about criticism of her work; the patient who sweats with anxiety in the dentist's office may be cool and courageous when scaling a sheer face of a mountain; and the high-flying business entrepreneur may avidly avoid taking social risks.

In 1968, I undertook a comprehensive review of the correlations that had been obtained in dozens of studies trying to link people's behavior in one situation (such as conscientiousness about meeting obligations and commitments at the office) to their behavior in another (such as conscientiousness at home). The findings shocked many psychologists. They revealed that although generally the correlations were not zero, they were much lower than had been assumed. The researchers who failed to demonstrate the consistency of behavior across different situations blamed their failure on having used imperfect and insufficiently reliable methods. I began to wonder whether the problem might be that their assumptions about the nature and consistency of human characteristics were wrong.

While the debate continued, it did not change the fact that across-the-board consistency of a person's behavior is generally too weak to be useful if the goal is to accurately predict from his behavior in one type of situation what he will do in another type of situation. Behavior is context-dependent. Highly developed self-control skills may be exercised in some situations and with some temptations, but not in others, as the stories of fallen public figures regularly remind us.

This raises problems in everyday life, and those problems became vivid for me when I needed to hire someone to take care of my young children for two weeks while I had to be abroad. I considered my neighbors' babysitter, "Cindy." She said she got good grades in high school, had worked as a lifeguard the previous summer, and did not smoke. She seemed to be a nice kid, and the neighbors agreed. But I also knew, as I described above, that we usually cannot predict accurately from one situation to other, very different situations. For example, how would Cindy behave at parties with her peers when the alcohol was passed around? We also couldn't predict, based on how she behaved babysitting one particular evening, how she would behave when babysitting my kids for two weeks straight. Yet that is how automatic impressions are formed. We compress bits of information into a compelling simplification, a stereotype that makes us feel that what is true in one situation is also true across other situations. Even highly confident, well-trained experts who often get it right also often get it wrong — especially when they try to predict specific behavior in different new situations.

I did not hire Cindy — she seemed too young — but I did hire a young couple who appeared mature and responsible. They made a very good impression on me during a long interview and visit to meet my children, who liked them. When I returned from the trip, however, I found the house transformed into a major mess, with ten days of unwashed everything waiting. The kids survived, but they were very unhappy and had developed an intense dislike for the couple — who had developed an equally strong dislike for them. My interest in doing more research on the consistency or inconsistency of behavior, particularly self-control and conscientiousness, intensified.

In time, my research team and I did find consistency, but not where it had been expected. We found it by looking closely at what different individuals did when we unobtrusively observed them, hour after hour, day after day, over the course of half a summer in a residential treatment camp for children. It was a natural laboratory for seeing in fine-grain detail how a person's behavior is expressed over time and across situations in everyday life, and it yielded some surprises that changed the understanding of personality. That story begins at Wediko.

15

IF-THEN SIGNATURES OF PERSONALITY

WEDIKO IS A WELL-ESTABLISHED summer residential treatment camp in a bucolic New England setting. When we conducted our research there, children aged 7 to 17 lived in rustic cabins for six weeks in small groups of same-sex peers similar in age, with about five adult counselors per cabin. The children were referred to the program because of significant social-adjustment problems in their home or school, particularly problems of aggression, withdrawal, and depression, and they came mostly from families in the Boston area. The goal of the camp's therapeutic environment was to enhance more adaptive and constructive social behavior.

My long-term research colleague Yuichi Shoda and I were allowed to do a large-scale research project at the camp in the mid-1980s, thanks to the staff at Wediko and Jack Wright, who was director of research at Wediko Children's Services. Jack, Yuichi, and the research staff systematically observed the behavior of the children over the course of the six weeks. The researchers extensively, but unobtrusively, recorded each child's everyday social interactions across a set of diverse camp activities and settings, from cabin time to time at the waterfront and in the dining hall, during arts and crafts, and so on. It was a massive data collection effort, and Yuichi and I collaborated with Jack on planning the project and analyzing the findings.

FINDING THE HOT SPOTS

The observers recorded what each child did during his or her interactions with others in the same set of situations, day after day, over the course of the summer. Jack, Yuichi, and I focused on an analysis of the hot-system negative behaviors — mostly verbal and physical aggression — that had brought the children to Wediko in the first place.

Strong emotions were not usually triggered when kids were stringing colorful beads or swimming, as long as all was going well. They flared up if one child deliberately wrecked the tower that another child had been working hard to construct, or if one responded to another's friendly invitation to work on the tower together by insulting him and cruelly making fun of him. To identify those "hot spot" psychological situations that triggered the children's aggression, the researchers first recorded how the campers and staff spontaneously talked about the children when they were asked to describe them. The youngest children qualified their characterizations quantitatively: Joe kicks and hits and yells — sometimes; Pete fights with everybody — all the time. Descriptions by the counselors and older campers, however, became more conditional when probed, and they were contextualized in particular kinds of emotion-arousing interpersonal situations — the "hot spots" that triggered the upset. "Joe always gets mad" would be the first declaration, but after a few more generalizations they would begin to specify the hot trigger situations: "*If* kids tease him about his glasses" or "*If* he gets a time-out."

Guided by these *If-Then* descriptions, the team observed what each child did, over and over again, during social interactions common in the Wediko camp environment. Five types of such situations were identified: three negative ("peer teased, provoked, or threatened," "adult warned," and "adult punished, giving 'time-out'") and two positive ("adult praised" and "peer approached pro-socially"). Each child's social behavior (for example, verbal aggression, physical aggression, withdrawal) was recorded as it occurred during each of the five situations. This provided an unprecedented sample of directly observed social interactions repeated often across the same set of situations, with an average of 167 hours of observation per child over the course of six weeks. These observations also made it possible to test two different predictions, reflecting different assumptions about human nature and how dispositions and behavioral tendencies are expressed: cross-situational consistency of personality and behavioral signatures of personality.

1. The classic and intuitively compelling conception of traits expects that in a given aspect of social behavior, like aggression or conscientiousness, people will consistently maintain their trait rank order across different kinds of situations. If we collect enough observations, we should be able to predict how they will

behave from one kind of situation to another. Going back to the fallen public figures in the headlines, we should expect a president who is conscientious in his public life also to be conscientious in his private life. Likewise, we should expect a highly aggressive child at Wediko to be highly aggressive across many different kinds of situations, with some individuals consistently higher and others consistently lower. That's called cross-situational consistency of personality.

2. In contrast, suppose our social behavior is generated not by stable broad traits expressed consistently across different situations but by our ability to make subtle discriminations based on how we interpret and perceive the different situations; the expectations and goals we bring to them; our past experiences in them; the emotions they stir in us; the competencies, plans, and skills we have for dealing with them; their importance and value to us; and so on. Then even the highly aggressive child will be *discriminatively* aggressive in some situations but not in others, depending on what the situation means to him. His hot system will make him predictably angry and his behavior explosive in the specific subset of situations that trigger his aggression — his distinctive hot spots. We call this situation-specific pattern of behavior the behavioral signature of personality.

"Jimmy" and "Anthony" are fictitious names for real children who participated in the Wediko study, and their behavior provides an example of what the research revealed. You can see Jimmy and Anthony's *If-Then* situation-behavior signatures across the five types of psychological situations we identify in the graphs that follow.

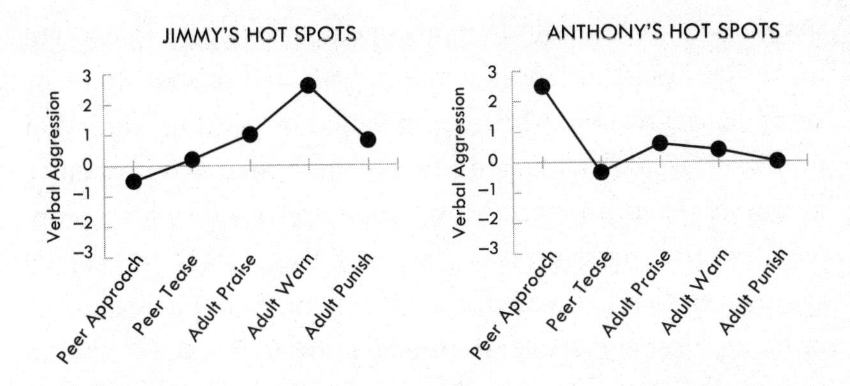

The graphs show each child's level of "verbal aggression" within each of the five situations over the course of the six weeks of camp. The horizontal line in each figure at zero shows the average degree of aggression within each of the situations that was exhibited by peers at Wediko that summer. The zigzag lines show Jimmy's (left) and Anthony's (right) distinctive patterns of deviation from that average. They reveal each child's unique hot spots: the specific situations in which their aggressiveness was significantly greater than that of their peers in the same situations. Verbal aggression, the behavior depicted here, is a euphemism for utterances like "You suck," "Your glasses are stupid," "You're a homo," plus the usual four-letter curses and associated gestures.

You see that Jimmy, in comparison with others, was exceptionally aggressive *If* warned or punished by adults. In fact, Jimmy was much more aggressive than his peers with adults no matter what they did, even when they were nice to him and praised him. When adults threatened him with impending punishment, he went wild. With his peers, however, Jimmy was not unusually aggressive, even when they teased and provoked him.

In contrast, Anthony was distinctively far more aggressive

than others at the camp *If* approached positively by a peer. For him, that kind of friendly overture was *the* hot spot that triggered his aggression, while being teased by peers or warned or punished by adults affected him no more than was normative among his Wediko peers. Most people, even if they have problems controlling aggression, are not likely to be especially aggressive when peers are friendly to them; Anthony is unusual in being most aggressive when people try to be nice to him, a formula for creating a miserable world for himself. He could not be more different from Jimmy, who was fine with peers but had a highly sensitive hot spot with adults regardless of whether they were punishing or praising him.

Although both boys are similar in their overall total level of aggression, their *If-Then* patterns reveal distinctively different hot spots. Once we recognize these, we can begin to think about what they mean and tell us about each person. Because *If-Then* patterns tend to be fairly stable across different situations with the same or similar hot triggers present, mapping them lets us predict future behavior in similar situations, identifies the individual's vulnerabilities, and can guide treatment and educational plans to better cope with them.

STABLE *IF-THEN* BEHAVIORAL SIGNATURES

Since the Wediko research, studies by other researchers with other populations and types of behavior have shown that stable *If-Then* patterns characterize most people when their behavior is closely examined. The behavioral signature of personality specifies what the individual does predictably *If* particular situational triggers occur. These behavioral signatures have been found with adults as well as children, and for everything from conscientiousness and sociability to anxiety and stress. Collectively, these findings contradict the classic and intuitively accepted trait assumption that people will be highly consistent across many different kinds of situations. What is stable and consistent is each individual's distinctive *If-Then* pattern. The patterns help us predict not just how much of a particular behavior trait a person will exhibit, but when and where he or she will behave that way. This information opens a window into what drives the behavior, and how it might change.

What we found for aggression at Wediko was also true for conscientiousness among students at Carleton College in Minnesota. That story began more than five years before the Wediko studies, when Philip K. Peake, friendly and smiling, sauntered into my office at Stanford in the fall of 1978. He had just completed his undergraduate studies at Carleton, wanted to work with me for a doctorate in psychology, and needed a place to securely store the many boxes, packed with fresh data, that he had brought with him. It was, and still is, the only time in my career that a student arrived not just with a good idea but with the enormous amount of data needed to test it. While still a college student, Phil had worked with Neil Lutsky, his adviser at Carleton, and had systematically tracked the behavior of Carleton College students across a set of different situations over the course of many months. He had assessed the students on different measures of "college conscientiousness" that had been selected by the students themselves. The measures ranged from class attendance to keeping appointments with instructors, returning reserve books to the library on time, room neatness, note-taking at lectures, and so on.

Just as we found at Wediko for aggression, there was little consistency in the Carleton students' conscientiousness across different situations. And their beliefs, indeed their intuitive convictions, that they were consistent were not related to their actual consistency across different types of situations. The same student who was consistently late for appointments with instructors might be highly conscientious about preparing carefully for exams weeks ahead of time. Then on what did they base those beliefs? Or were they just illusions of consistency? Their beliefs turned out to be linked — very strongly — to their *If-Then* conscientiousness patterns: the more repeated or stable these patterns were over time, the more the students felt that they were consistently conscientious across different situations. They believed they were consistent because they knew their predictable long-standing *If-Then* behavior signatures. A Carleton College student thought he was consistent in his conscientiousness at school because he knew, for example, that he was always punctual for class and for appointments with professors, and he also knew that his room and his notes were always a mess, and he was invariably late with homework assignments. It is the stability of our *If-Then* patterns over time that leads us to think that we consistently exhibit a particular trait. Our intuition of consistency is neither paradoxical nor illusory. It is just not the kind of consistency that researchers had been looking for during much of the last century. That is useful to know because it tells us where to look if we want to predict what other people will do — as well as what we ourselves are likely to do.

These findings madc it casicr for me to answer reporters' questions about whether or not President Clinton was trustworthy. Nighttime behavior with White House interns in the Oval Office should not be expected to predict conscientiousness and responsibility regarding a president's behavior when he deals with heads of state to negotiate an agreement in the Rose Garden the next morning. When asked "Who is the *real* Bill Clinton?" my long answer was that he is highly conscientious and self-controlled in some contexts, but not in others; both sides of him are real. If you want to add up all his conscientious behaviors regardless of the context, he will, on average, be highly conscientious — although how high depends on whom you compare him with. And how you decide to evaluate his overall behavior, as well as whether or not you like or respect his *If-Then* patterns, depends on you.

MAPPING HOT SPOTS: *IF-THEN* STRESS SIGNATURES

If you draw a map of what triggers your hot system, you might be surprised. A map of your *If-Then* situation-behavior signatures can alert you to your hot spots and when and where you are prone to react in ways that you are likely to later regret. Self-monitoring to discover these hot spots can become a step toward reappraising those situations and cooling them, giving you more control over your behavior in pursuit of the goals and values that matter most to you. Even if you don't want to cool those automatic reactions, you might still benefit from tracking them and observing their consequences.

In one study, adults who were suffering from high stress were instructed to use the *If-Then* assessment approach to find the hot spots that triggered their stress. In carefully structured daily diaries, they kept track of the specific psychological situations that activated high stress for them, and they described their reactions to each of those hot triggers day after day. For example, "Jenny" had normal stress levels on average, across diverse situations — indeed, she was somewhat below average. Her problematic stress signature emerged only when she felt socially excluded; in those situations her stress level escalated. When she felt excluded, she became distressed, blamed herself, blamed others even more, and became avoidant. Helping Jenny discover the psychological situations in which she did and did not experience stress, and also helping her identify her reactions in those situations, was the first step in devising a targeted intervention to enable her to deal more adaptively with them. While this study focused on *If-Then* stress signatures, the same self-monitoring in a diary or tracking device can be used to map triggers for overreactions resulting from any feelings or behaviors of concern. Once you know the *If* stimuli and situations that trigger behaviors that you want to modify, you are positioned to change how you appraise and react to them.

SELF-CONTROL COOLS AGGRESSIVE TENDENCIES

The highly aggressive behavior patterns that brought children to Wediko for treatment not only put the individuals at high risk for problem-filled lives but also created risk for others to become potential victims of uncontrolled aggression. I discussed in earlier chapters how self-control ability can have a protective effect, for example against the destructive effects of severe rejection anxiety. Might self-control ability also help control the expression of strong aggressive tendencies?

The Wediko research gave us an opportunity to test this possibility. In a study led by my then postdoc Monica Rodriguez (now a professor at the State University of New York in Albany), the Wediko children took a version of the Marshmallow Test in which the treats were M&M's candies — a few now or a bigger bag later. Some of the children spontaneously used a cooling strategy to reduce their frustration; they avoided looking at the candies and the bell and purposefully distracted themselves from the temptation. Like the other Wediko children, these self-distracters were at risk for problems of uncontrolled aggression. But they were much less physically and verbally aggressive over the course of the summer compared to those who did not use a cooling strategy while trying to get more candies. The same cognitive skills and executive function that allowed them to distract themselves on the test seemed to help them cool and control their aggressive reactions when their hot spots were activated in interpersonal conflicts at the camp. In the split second between starting to become angry and violently striking out they managed to cool down just enough to stop themselves from losing control completely.

Regardless of how good one is at self-control, however, there are situations that can undo willpower and traumatize well-functioning people in seemingly irrational, indeed maddening, ways.

16

THE PARALYZED WILL

JOHN CHEEVER'S 1961 SHORT STORY "The Angel of the Bridge" shows us how easily the cool system can be crippled, even when self-control skills are excellent, the psychological immune system is doing its best, and the motivation to exert self-control and willpower could not be stronger. Cheever's protagonist is a successful businessman who lives in Manhattan, and one evening, as he approaches the George Washington Bridge to get home, a violent thunderstorm suddenly strikes full force. The wind is ferocious, the huge bridge feels as if it is swaying, and our nameless protagonist (let's call him Bridgeman) panics at the terrifying thought that the bridge will collapse. He manages to make it back home, but he soon discovers that he has developed an incapacitating fear not just of that particular bridge but of other bridges as well. Bridgeman's work requires him to frequently traverse bridges, and he desperately tries to use willpower to get over his fear — but all his efforts fail and he becomes increasingly depressed, ruminating that he is spiraling out of control.

HOT CONNECTIONS

When Bridgeman was crossing the George Washington Bridge as the storm suddenly hit, the structure that he had calmly crossed so often in the past changed its emotional impact for him; overwhelmed by stress, his hot system automatically associated the previously neutral bridge with the terrifying emotional experience that he had on it when he felt it swaying. He panicked, thinking that it was splitting, and he imagined himself being hurled into the turbulent waters below. When a neutral stimulus, like a strong and beautiful bridge, becomes associated in the hot system with an intense fear experience, the fear can easily generalize to many other related but previously neutral stimuli — in this case, to other big bridges high above water. Within the hot memories of his amygdala, even the thought of crossing a large bridge subsequently reactivated his panic in the storm. No matter how hard Bridgeman tried to use his cool system to reappraise the experience, to exert willpower, to rethink his plight, to self-distance and gain perspective, it was impossible for him to get over it by exerting sheer will.

When such hot connections form between an innate fear response and a previously neutral stimulus, we can become as helpless as the dogs in the laboratory studies of "classical fear conditioning" that took place early in the last century. The unfortunate dogs received an electric shock every time a buzzer sounded, and they soon became the buzzer's emotional victims: even when it no longer signaled electric shock, they still became terrified. Willpower and cooling skills do not help people overcome this kind of collateral damage. Bridgeman's bridge-crossing behavior was no longer within his control; it was under stimulus control, ruled automatically and reflexively by his hot system. Consequently, all his efforts to exert willpower and be tougher failed, making him increasingly desperate, even fearing that he was losing his mind.

Happily, in Cheever's story, an "angel" rescued Bridgeman. It happened one sunny day when he could not find a bridge-free route to his destination, and as he neared the bridge he had to cross, his terror returned. Unable to go on, he had to pull to the side of the road. Then a lovely angel-like young girl, carrying a small harp, approached and asked to hitch a ride with him. As she serenaded him across the long bridge with a sweet folk song, his fear dissolved. Bridgeman still remained careful to avoid the George Washington Bridge, but crossing other bridges soon became routine once again.

Cheever's story anticipated cognitive behavior therapy by many years and flew in the face of the then-dominant medical disease model of psychological problems. In the medical model, it is essential for the physician to separate the presenting complaint and its potential cause, and then illuminate the cause. For example, for a patient with symptoms of back pain caused by a cancerous tumor, prescribing painkillers rather than removing the cancer of course would soon prove disastrous. But for psychological conditions that cripple the individual, the presenting complaint — like the terrifying fear of bridges — often *is* the problem that has to be addressed and removed.

The belief that the medical model of illness also applied to phobias was widely shared for many years. The prevalent worry about simply treating the behavior problem, the "symptom," was that this would lead to substitution of another symptom and much worse problems. It was assumed that the underlying causes had to be in early childhood traumas of which the individual was unconscious, and that these causes had to be uncovered with a long analysis.

REWIRING THE CONNECTIONS

In 1958 Joseph Wolpe, a psychiatrist who became skeptical about psychoanalytic theory, took the risk of attempting direct behavior modification with patients who suffered from anxiety and panic attacks like those in Cheever's story. Wolpe proposed: "If a response antagonistic to anxiety can be made to occur in the presence of anxiety-evoking stimuli so that it is accompanied by a complete or partial suppression of the anxiety responses, the bond between these stimuli and the anxiety responses will be weakened."

Wolpe thought that deep muscle and breathing relaxation exercises could help patients develop the needed antagonistic responses to anxiety, and then slowly the relaxation response would become connected to the feared stimulus until the fear dissipated. In this type of therapy, the relaxation response is at first connected to stimuli that are only remotely related to the traumatic stimulus (e.g., pictures of small bridges over calm, shallow ponds bathed in sunshine). Then, step by step, as anxiety about these milder versions of the threat is overcome, the patient moves to the next, more fearful representation of the stimulus — until finally the relaxation response is linked to thinking about, and ultimately actually approaching, the feared stimulus itself. And at this point, if it is the George Washington Bridge, the patient can cross it in a relaxed state. As Cheever's story suggests, this slow process sometimes can be dramatically shortened when the relaxing event, antagonistic to anxiety, arrives full blown in the form of a lovely angel who sings you across the bridge, more likely in fiction than in life.

Cheever's story was a preview of what soon became the standard approach for overcoming all sorts of phobias, without having to wait for an angel. In many studies the panicked person was put into a safe situation in which he or she observed bold models who slowly but fearlessly, step by step, approached the feared stimulus and demonstrated that they could remain calm and unharmed. At about the same time that the marshmallow experiments were being done at Stanford, Albert Bandura, my colleague while I was there for more than twenty years, was doing studies of preschool children who had become terrified of dogs. From a safe distance the preschoolers observed as a model fearlessly approached a dog. At first the model (a graduate student assisting in the research) just petted the dog a bit while the animal remained confined in a playpen, and then gradually she joined the dog in the playpen, affectionately hugging it and feeding it treats. The children who were watching overcame their fears rapidly and soon were hugging and petting dogs themselves. Bandura and his colleagues achieved similar results even more economically for a wide variety of fears in both children and adults by exposing the frightened participants to fearless models acting out scenarios on films. These studies became important foundations for the treatment of fears in cognitive behavior therapies.

Bandura's research showed that the best way to overcome phobias is to first observe the fearless model and then, with the model's guidance and support, try it and master it yourself. Using a variety of "guided mastery experiences," both children and adults overcame not just fears of dogs, snakes, spiders, and so on but even the most profound and disabling anxiety disorder, agoraphobia: the fear of going outdoors. Discussing his research, Bandura commented that some of the phobic people studied had been plagued by recurrent nightmares for decades, but the guided mastery treatments even transformed their dreams: "As one woman gained mastery over her snake phobia, she dreamt that the boa constrictor befriended her and was helping her to wash the dishes. Reptiles soon faded from her dreams. The changes endured. The people with phobias who had achieved only partial improvement with alternative modes of therapy achieved full recovery with the benefit of the guided mastery treatment regardless of the severity of their phobic dysfunctions."

The popular 2010 film *The King's Speech* showed the effectiveness of direct behavior modification for helping the man who became King George VI of Britain overcome his distressing speech impediment. When His Majesty conquered his stammer, he was able to become the strong monarch his nation needed in a time of war. His sense of self-worth and his personal life flourished; losing the stammer, regardless of its root cause, produced only gains — no deficits, no replacement costs.

Thirty years after the king lost his stammer, in a much less dramatic but well-controlled and compelling experiment, the psychologist Gordon Paul assigned college students who dreaded public speaking to different conditions. In one group, they learned a desensitization procedure to systematically relax deeply while imagining situations connected to public speaking. They learned to remain relaxed as the situations became increasingly threatening, from reading about speeches alone in their room, to getting dressed the morning on which they had to give a speech, to presenting a speech before an audience. Another group received brief insight-oriented psychotherapy from an expert clinician to explore the possible reasons for their anxiety. Still another group obtained placebo "tranquilizers" that supposedly would help them deal with the stress. The clear winner on all measures, from ratings of their anxiety when speaking to physiological measures of their anxiety, was the group that learned the desensitization procedure. The students in this condition not only overcame their fear of public speaking but also significantly improved their overall college grades. Helping people directly overcome problems like their speech disorders, or irrational fears, or facial twitches — which may or may not be symptoms of other problems — does not create worse problems. When done right, it makes them feel better about themselves and improves the quality of their lives.

It took many decades and studies like these to finally overcome the worries that earlier therapists had about symptom substitution and to at last develop an evidence-based, economical treatment for helping people get over unfortunate hot-system associations. Cognitive behavior therapy is now, for the most part, standard practice in the United States. In many parts of the world, however, it has not been accepted, or at best is considered insufficient. I recently told a good friend, a practicing clinician who works with disturbed children, about "The Angel of the Bridge," thinking she might find the story useful in her work. She smiled, shrugged her shoulders, and dismissed it as a superficial treatment — a palliative, as misguided as prescribing sedatives to treat cancer. My clinician friend believes the bridge fear to be merely an expression of a deep underlying anxiety. She is convinced that when the bridge fear is removed it will be replaced by worse symptoms, because the underlying anxiety that caused it is buried by repression in the hot system. You need lengthy analysis, she argued, to get to the bottom of it. When I asked her how she would proceed if Bridgeman were her patient, her answer was quick. She noted that Bridgeman's fear was really about falling into the existential void, and the treatment would have to address that deeper fear and its possible roots. I was impressed by how poetic her answer was, but doubtful that it would help get Bridgeman across the George Washington Bridge.

Bridgeman's dilemma illustrates how difficult it can be, even for someone usually good at self-control, to overcome the hot system's automatic associations. To summarize, those associations can instantly and reflexively connect intense emotional reactions (especially fear) triggered by the amygdala to the stimuli that were present when the fear-producing event occurred, even though those stimuli were emotionally neutral to begin with. Overcoming this collateral damage, created accidentally, requires rewiring the connections. The fear produced when Bridgeman felt that the bridge was about to split during the sudden storm had to be disconnected from bridges. Neither Bridgeman nor anybody else may be able to do this alone, but a first step is to understand how these fear-producing associations form and how they can be overcome. The goal becomes, in Bridgeman's problem, disassociating bridges from fear and reconnecting them with the pleasure of safely crossing to get to the other side. Absent an angel and a harp, or even a therapist, a friend could drive the fearful person first over very short bridges suspended just a few feet above shallow water, and then maybe before the day ends, over larger, higher ones, perhaps while lovely harp music plays on the radio. The friend might then sit next to the fearful person while he takes the wheel and tries to drive first over the little bridges that span nearly dry land, and then gradually over the bigger ones high up above the water, as each bridge begins to feel safe again. This kind of desensitization allows us to escape stimulus control and restore self-control. It can free the paralyzed will.

17

WILL FATIGUE

THE EXHAUSTED AUDIENCE WAS waiting for the program to start at the reception in the elegant Hungarian consulate on the Upper East Side of New York. It was late in the evening after a long workday, with a "patron of the arts" crowd, age fortyish and much above, most in business attire, gray or black, reexamining Rolex watches and iPhones, eyes starting to close. After a long delay, amplified music suddenly burst forth, full blast:

> *I WANT TO DO THE BAD THING NOW! AND I DON'T MIND SUFFERING LATER!*

The ragtag band on the stage screamed the words exuberantly, wildly playing fiddles and guitars, banging their drums and metal cans, shaking castanets and rattles, wearing tiny old fedora hats and hippie clothes, and flirting outrageously with one another and the very staid audience to promote tourism for Hungary. It electrified the dozing crowd, eliciting the excited cheering and roaring that might come from kids at a rock concert. If instead the program had begun with the expected video and lecture on the wonders of Budapest, the forced coughing fits and escapes toward the exit would soon have followed.

Before the band stirred them up, the audience seemed to be having a bad case of collective will fatigue, tired from their own excessive exertion of self-control. Everyday willpower efforts, just to make it through a long and stressful workday, can be exhausting. They were primed to let the grasshoppers in them have some pleasure, *right now,* and they were thrilled to accept the band's invitation to cut loose, have fun, and let the hot system enjoy, while the overworked cool system took a break.

THE TIRED WILL

Are there limits to how much self-control and delay of gratification we can exert before will fatigue takes over? The concept of a fatigued will that becomes drained by its own excessive use is the basic idea underlying a current influential scientific theory on the nature of willpower and self-control. And it has important implications for how you think about your own ability to self-regulate.

Roy Baumeister and his colleagues see willpower as a vital but limited biological resource that can easily be depleted for temporary periods. Their "strength model of self-control" proposes that self-control depends on some internal capacity that relies on a limited amount of energy. This is much like the traditional concept of "the will" as a fixed entity or essence: some people have a lot of it, others very little. According to this model, self-control is like a muscle: when you actively exert volitional effort, "ego depletion" occurs, and the muscle soon becomes fatigued. Consequently, your willpower and ability to override impulsive behavior will temporarily diminish on a wide variety of tasks that demand self-control. This could affect everything from mental and physical endurance to rational thinking and problem solving, from response inhibition and emotion suppression to making good versus bad choices.

Suppose that you are famished and eager for a snack at the annual office reception. If you manage to forgo the tempting, freshly baked chocolate chip cookies in front of you, and instead make yourself stick to just the vegetable tray, the strength model suggests that immediately after this you will expend less effort on unrelated tasks that continue to require self-control. Evidence for this idea surfaced in a classic experiment that has become the prototype for studying ego depletion. College students taking introductory psychology at Case Western Reserve University in Ohio were required to participate in psychology experiments as part of their course, and those who went to Professor Baumeister's laboratory for their course requirement were put into the Radish Experiment. The students arrived hungry because they had been told to fast before coming. Once in the lab, they were asked to force themselves to forgo the tempting chocolate chip cookies and candy and eat some radishes instead. Right after that they were asked to work on geometry problems that were actually impossible to solve. The study showed that they quit much sooner than the students who had been allowed to eat the cookies and candy.

In more than a hundred other experiments, researchers demonstrated similar results: engaging in self-control at Time 1 reduced self-control at Time 2, which immediately followed Time 1. This was true no matter which act of self-control the students were instructed to perform. The results were the same, whether they were stifling their emotional reactions to an intensely emotion-provoking film about what happens to wildlife in a nuclear wasteland (A *Dog's World*, originally titled *Monde Cane*), or avoiding thoughts of white bears once primed to think of them (try it if it sounds easy), or reacting kindly to a partner's bad behavior.

MIND OVER MUSCLE

Students did indeed reduce their subsequent efforts in many studies like these, but later research showed that the reduced efforts were probably not caused by the reasons the researchers had initially assumed. As the demands for effortful self-control and tedious work escalated, but the incentives did not, the students' attention and motivation shifted. Rather than having their willpower "muscles" depleted, they probably became fed up, feeling that they had complied sufficiently with the experimenter's demands to do boring tasks. In one task, for example, after spending five minutes crossing out every "e" in a typewritten text, students then had to *not* cross out an "e" if it was followed by a vowel. And when people are given strong incentives to persist even on tasks like that, they do continue longer. As motivation to exert self-control increases, effort continues. With no increase in motivation, it does not. In this interpretation, the reduction in self-control is not due to a loss of resources: it reflects, instead, changes in motivation and attention.

The feeling of exhaustion, of being "done in" by effortful work, is real and anything but rare. Yet we also know that when sufficiently motivated, we can keep right on going — sometimes even with increasing zeal. When in love, we can go from experiencing an exhausting day or week or month to running eagerly to wherever our beloved is. For some people, feelings of fatigue become the cues not for turning on the television but for jogging to the gym. The motivational interpretation of effortful persistence simply argues that mind-sets, self-standards, and goals guide when we become invigorated rather than drained by our efforts, and when we need to relax, nap, self-reward, and allow the grasshoppers in us to emerge.

If you believe that persisting on tough tasks is energizing rather than depleting, will it protect you from will fatigue? Indeed yes: when people are led to think that effortful tasks will invigorate rather than drain them, they improve their performance on a later task. For example, when people were primed to believe that they would be energized by controlling their facial expressions (so as not to show the emotions they were experiencing), they performed better on the later task of squeezing a handgrip. Their later performance was not impaired by the earlier effort, and their egos did not deplete.

At Stanford University, Carol Dweck and her colleagues found that those who believed that their stamina fueled itself after tough mental exertion did not show diminished self-control after a depleting experience. In contrast, those who believed that their energy was depleted after a strenuous experience did show diminished self-control and had to rest to refuel.

Dweck's team went on to track college students across three points in time, the last being their final exam period, which demanded strong self-regulation. Students who had an implicit theory of willpower as a non-limited resource fared much better during the high-stress exam period than those with a limited-resource theory, who reported eating more unhealthy food, procrastinating more, and ineffectively regulating while trying to prepare for their tests. These findings underscore the importance of how we think about ourselves and our capacities for control, and they undermine the idea that our ability to exert effort in pursuit of goals is an immutable, biologically driven process.

WHEN YOU CONTROL THE TREATS: SELF-REWARD STANDARDS

It does not take experiments or philosophers to know that an excess of will can be as self-defeating as its absence. Always postponing gratification and continuously working and waiting for more marshmallows can be the unwise choice. When the world is full of uncontrolled inflation, bank failures, and promised future payoffs that never deliver, there are good objective reasons to ring the bell and refuse to wait. And the subjective reasons are just as compelling. In the extreme, delay of gratification becomes stifling, a joyless driven life of postponed pleasures, happy diversions not taken, emotions not experienced, possible lives unlived. We are both ants and grasshoppers, and to lose the hot emotional system and live continually dominated by the cool cognitive system in the service of a possible future can become a life story as unsatisfying as its opposite.

When do we feel entitled to behave more like grasshoppers than like future-oriented ants constantly busy with work? When do we allow ourselves to relax, let the hot system take over, self-reward with our viscerally preferred personal marshmallows, and forget the unanswered emails and tomorrow's to-do list?

What determines our willingness to let ourselves have the pleasure of doing nothing, the unscheduled weekend at the beach, the trip to the big city, or just time off at home to celebrate life? We may not need to act as stupidly as some of those fallen heroes in the headlines, but we all do seem to have implicit rules about when we suspend self-control and let ourselves enjoy the fun thing *now*, or instead postpone those pleasures and keep pushing on for more and bigger rewards in the future. How do we develop those rules? Answers to these questions have direct implications for how we raise our children and how we treat ourselves.

Today, upper-middle-class American parents supposedly live child-centered lives, rushing home from work to assure maximum "quality time" devoted to the kids, showering them with affection and rewards, letting them lead the way. They can often be seen allowing their children to have unchecked screaming fits because it takes a few minutes for their hamburger to arrive at McDonald's. In contrast, French parenting is reputed to raise preschoolers who can be taken to elegant restaurants in Paris where they ostensibly sit quietly and wait for their entrecôte with haricots verts while their parents enjoy an aperitif. To raise ideal kids, one Chinese American mother offers a long list of what should be forbidden, including sleepovers, play dates, TV, computer games, and any grade lower than A. That is the formula in 2011 that Amy Chua proposed in her *Battle Hymn of the Tiger Mother* to raise a child who's likely to excel at playing either the violin or the piano and be number one in every class (with the possible exception of gym).

A dozen years earlier, Judith Rich Harris argued that parenting of any kind does not really matter anyway because socialization by peers and genetics are the two key factors that shape children's lives. To go beyond anecdotes and personal opinions, we would have to conduct experiments that carefully manipulate what happens under different parenting conditions in real life, but such studies cannot be done. We can ask and answer questions relevant to parenting practices, however, by doing short-term experiments with adult models under realistic conditions that are meaningful to children.

My interest in this area began when my children were attending elementary school and brought home their proudest early achievements, like the blue and black flip-flop sandal made out of baked clay by my youngest daughter. This led me into a series of studies to see how we set standards for our own accomplishments starting early in life, and how we do or don't reward ourselves when those standards are met. The questions became: What are the socializing experiences and the implicit rules that guide this form of self-reward and self-regulation? When do children develop will fatigue and decide that it is time to congratulate themselves, indulge a bit, and reward themselves? When do they persist and delay gratification until they meet more stringent standards? Or does the continuing effort itself become the pleasure?

MODELING SELF-STANDARDS

Because models profoundly influence who we become, I was eager to study how they guide the standards we develop for evaluating and regulating ourselves beginning in childhood. The characteristics and behaviors of adult models influence what young children learn, imitate, and transmit to others. At Stanford, concurrent with the marshmallow studies, my students and I began to do experiments to see how children acquired their self-standards. In these studies, we varied the model's attributes and self-reward behavior to see how they influenced what young children incorporated into their own standards when the adult left the room.

My student Robert Liebert and I selected fourth-grade boys and girls, mostly ten-year-olds, from local elementary schools near Stanford. In individual sessions, we introduced each child to a young woman (the model) who showed him or her "a sort of bowling game" that a toy company was ostensibly testing to see how much children liked it. It was a miniature, three-foot-long version of a bowling alley, with signal lights at the end that registered the score for each trial. The target area at the end of the runway was screened so that the bowler could not see where the ball hit and relied on the score displayed in signal lights for feedback. These scores were preset and not connected to actual performance, but in a way that made them completely credible. Within easy arm's reach was a large bowl full of tokens — colorful poker chips — that the child and the model could use to reward themselves for their performance. They were told that the chips were worth valuable prizes at the end, and the more chips, the better the prize. The attractively wrapped prizes were in full view in the room but were not discussed.

DO AS I SAY OR DO AS I DO?

To play the game, the model and the child took turns, one trial at a time. In order to simulate different parenting styles, we created three different scenarios for how the model rewarded her own performance and how she guided the child to evaluate and reward his own performance. Each child participated in only one of these conditions.

In the "tough standards" scenario, the model was stringent with herself and equally stringent with the child. She took a token only when her score was very high (20), making self-approving comments like "That's a good score; that deserves a chip" or "I can be proud of that score; I should treat myself for that." Whenever her score was lower than 20, she refrained from taking a token and criticized herself (e.g., "That's not a very good score; that doesn't deserve a chip"). She treated the child's performance in a directly parallel way, praising the high scores but remaining critical of lower scores. In the "tough on model, easy on child" script, the model was tough on herself but lenient with the child, leading him to self-reward for lower scores. In the "easy on model, tough on child" scenario, she was lenient with herself but held the child to a stringent standard of self-rewarding for only the best score.

After the children participated in one of these conditions, we unobtrusively observed their spontaneous self-reward behavior when they bowled alone in the post-test in which the tokens remained freely available. Children adopted the most stringent standards for self-reward when they had learned from a tough-on-herself model who was equally tough on them. This model encouraged them to reward themselves only for top scores and held herself to the same standard. When the modeled and imposed standards were consistent, children adopted those standards without a single deviation in the model's absence, in spite of the stringency of the criterion and the desirability of the rewards. The research also showed that these effects were especially strong when children believed that the model was powerful and had control over many highly desirable treats and rewards.

Children who were encouraged to be lenient with themselves remained that way in the post-test when they were left on their own, even if they had observed a model who was stringent with herself. In the group of children who were held to a stringent self-reward standard during training but had learned from a model who was lenient with herself, half retained the more stringent standards that had been taught, and half used the more liberal standards they had observed the model use for herself. This study suggests that if you want your children to adopt high self-reward standards, it's a good idea to guide them to adopt those standards and also model them in your own behavior. If you aren't consistent and are tough on your children but lenient with yourself, there is a good chance they'll adopt the self-reward standards you modeled, not the ones you imposed on them.

MOTIVATION AND EFFORT: THE GREEN TEAM

If we step outside the laboratory, we can look at the psychological conditions and human qualities that motivate people to push themselves to the extremes of self-control, a prime example being the United States Navy SEALs. In his 2012 autobiography, Mark Owen (a pseudonym) describes the raid in which he and his teammates killed Osama bin Laden, and it is a thriller that goes beyond the excitement of the raid to the motivations and training that help shape individuals like Mark to defy will fatigue.

Mark was the child of missionary parents in Alaska. In junior high school, he opened *Men in Green Faces*, a book by a former SEAL. It depicted the SEAL firefights and ambushes in Vietnam's Mekong Delta, particularly focusing on their hunt for a rogue North Vietnamese colonel. Mark was captivated from the first page, immediately sure he wanted to be a SEAL: "The more I read, the more I wanted to see if I could measure up. In the surf of the Pacific Ocean during training, I found other men just like me: men who feared failure and were driven to be the best. I was privileged to serve with and be inspired by these men every day."

SEAL training was brutal, involving endless long runs in freezing temperatures or desert heat, extreme physical challenges like pushing cars and buses, and searching and shooting in kill houses under realistic and endlessly unpredictable battle conditions. For people like Mark, reaching the hundredth pull-up became the signal for raising the bar and making the next thirty the goal; beating their own personal best became the moving target, the self-standard that they always tried to exceed — not the signal to feel fatigue and let themselves quit. In a program in which 75 percent of the men in each class fail to complete training, Mark finally made it to the Green Team, the last step toward becoming prepared and selected — maybe — for the elite Navy SEAL Team Six, which conducts the most dangerous and difficult hunt and kill missions. If selected, Mark would fulfill his burning lifelong goal.

Mark's experiences and triumphs illustrate the importance of an implicit theory of willpower that is open to virtually limitless development, combined with burning goals that fuel and sustain effort and grit, and a social environment that provides inspiring models and support. All of these play into the relentless training and self-discipline needed to become truly exceptional — whether the goal is to play Bach at Carnegie Hall, win the Nobel Prize in physics, gather gold medals at the Olympics, move from poverty in the South Bronx to Yale University, become a Navy SEAL, or, in the preschool version, collect marshmallows — when 15 minutes feels like a lifetime.

PART III

FROM LAB TO LIFE

I BEGAN PART I with the story of the Marshmallow Test and the experiments that revealed the strategies preschoolers used to control themselves. Part II showed that the same strategies also empower adults to postpone pleasures to save for their retirement. In Part II, I also illustrated that the same mechanisms underlying successful control strategies help heartbroken people overcome their pain, rejection-sensitive individuals preserve their relationships, and exhausted Navy SEALs do even more pull-ups. Taken together, what has been discovered about mastering self-control leads to several key conclusions:

1. Least surprising, some people are better than others in their ability to resist temptations and to regulate painful emotions.

2. More surprising, these differences become visible as early as in the preschool years, are stable over time for most but not all people, and predict highly consequential psychological and biological outcomes over the course of life.

3. The traditional belief that willpower is an inborn trait that you either have a lot of or you don't (but cannot do much about it either way) is false. Instead, self-control skills, both cognitive and emotional, can be learned, enhanced, and harnessed so that they become automatically activated when you need them. This is easier for some people because emotionally hot rewards and temptations are not as hot for them, and they also more readily cool them. But no matter how good or bad we are at self-control "naturally," we can improve our self-control skills and help our children do the same. Moreover, we can fail to develop our self-control skills, and even if we have them in abundance, we may lack the goals, values, and social support needed to use them constructively.

4. We don't have to be the victims of our social and biological histories. Self-control skills can protect us against our own vulnerabilities; they may not eliminate these vulnerabilities completely, but they can help us function better with them. For example, an individual with high rejection sensitivity who also has good self-control is better able to protect the very relationships that he fears he will lose.

5. Self-control involves more than determination; it requires strategies and insights, as well as goals and motivation, to make willpower easier to develop and persistence (often called grit) rewarding in its own right.

In Part III, I turn from the lab to life, looking first at how these findings speak directly to public policy. Then I summarize and illustrate the core strategies that can make willpower in daily life less effortful and more natural for our children and ourselves. In the final chapter, "Human Nature," I discuss how the research about self-control and the plasticity of the human brain changes the conception of who we are.

18

MARSHMALLOWS AND PUBLIC POLICY

MANY YEARS AGO, WHEN I was a student in the clinical psychology graduate program at the City College of New York, I worked as an uncredentialed social worker with groups of economically impoverished children and adolescents. I met with them at the Henry Street Settlement, an agency in "the slums," as they were then called, of the Lower East Side of Manhattan. I was intrigued by the classic clinical psychology theories and methods I was learning at school and was eager to apply them in my social work.

One evening at Henry Street, I was surrounded by a group of adolescent boys who listened while I tried to use my new insights to interpret the anger of a particularly hostile youngster, a boy whose older brother was awaiting execution in the state prison's death row. The kids seemed particularly attentive and eager to learn more, but I soon smelled smoke and saw that the back of my jacket had been set on fire by one of the boys behind me.

After putting it out, I recognized that the fascinating clinical methods and concepts I was being taught were irrelevant, to say the least, to the young people I was supposed to help. This insight was one of the steps that led me to a research career, as I hoped to find more effective ways to help kids like those at Henry Street make the most of their lives.

Half a century later, I started to hear from educators who were trying to apply the findings from research on self-control and delay of gratification to the immense challenges they were facing, as the gap kept widening between those at the top of the economic and achievement ladder in the United States and those at the bottom. While much public education continues to deteriorate, it is always encouraging to meet dedicated and creative educational leaders who are crafting alternatives. It is a privilege to glimpse what they are doing, learn of the innovations they are trying, and see their successes, frustrations, and challenges. Their commitment to nurturing the qualities essential for success in their students, and their eagerness to try to apply the research findings in their daily efforts, helped motivate me to write this book. In this chapter I look at how the discoveries from the research on self-control can be incorporated into educational interventions and the resulting implications for public policy.

PLASTICITY: THE EDUCABLE HUMAN BRAIN

A silent revolution in the conception of human nature has been slowly building momentum over the past two decades, as scientists reveal the plasticity of the human brain. The unexpected finding is that there is great malleability in the areas in the prefrontal cortex that enable executive function. As discussed throughout this book, these mechanisms allow us to cool and restrain our impulsive hot reactions in the service of our goals and values, and to regulate emotions adaptively.

The importance of executive function (EF) for how lives play out, and specifically for our ability to overcome stimulus control with self-control, is undisputed. The public policy implications that follow depend on whether or not we think that EF skills and the potential for self-control are essentially prewired and fixed. If they are, there is little that interventions can do. But if they are malleable, the public policy implications are profound and call for educational efforts to target the enhancement of these skills as early in life as possible.

We know today that when a preschooler manages to wait for her two marshmallows, the anterior cingulate and lateral prefrontal areas of her brain must activate strongly. These areas are key parts of the cool cognitive system she needs to control the impulsiveness of her emotional hot system. Imaging with fMRI was still decades away when I was watching the children through the observation window, and back then I could not begin to imagine what went on in their brains as they sat facing their treats in the Surprise Room. Well-controlled laboratory interventions since then reveal that direct training of EF yields not just improvements in self-control but also changes in the corresponding neural functions in the brain.

In 2005, a research team under the leadership of Michael Posner conducted experiments to show how training and genetics jointly influence the cognitive and attention-control skills that let preschoolers cool their hot systems. The researchers exposed children aged four to six to a 40-minute attention-training session each day for five days. In these sessions, the children played a variety of computer games designed to tap and enhance different aspects of their attention-control ability — in particular, the ability to keep a goal in mind and shift their attention to pursue it while inhibiting interfering impulses. In one game, for example, they used a joystick to track a cartoon cat on the computer screen. Their job was to move the cat to a grassy area and avoid the muddy areas, which kept getting bigger while the grassy areas began to shrink, making their task increasingly difficult.

The question that the researchers were trying to answer was: will such training experience influence the children's attention-control scores later on a different standard test of attention control? Their attention control did improve significantly when compared to a no-training control group — an encouraging finding given the simplicity and short duration of the training. Most surprising was that even this brief training period served to improve scores on non-verbal measures of intelligence.

The same group of researchers went on to find in related studies that specific genes that influence the child's ability to cool and control negative emotions and reduce hyperactivity also influence attention and self-control ability. The DAT1 gene in particular has a role in various dopamine-related disorders, including ADHD, bipolar disorder, clinical depression, and alcoholism. The promising news for public policy is that the researchers discovered that even in people with the genetic vulnerability, attention control can be enhanced significantly by interventions, specifically through better education and parenting techniques during development. It is nature *and* nurture seamlessly influencing each other.

Given the importance of EF for developing social as well as cognitive skills and self-control, it is good to see the research by Adele Diamond at the University of British Columbia that tests to determine if EF is in fact malleable and teachable in simple educational interventions in preschool. In 2007, Diamond and her colleagues reported the results of one of her largest studies in the journal *Science*. Their Tools of the Mind curriculum, designed to enhance EF development, exposed preschoolers (average age of 5.1 years) intensively and daily to forty EF-promoting activities. These ranged from gamelike exercises in which the child tells herself what she should do, to dramatic play, to practicing simple tasks that improve memory, to learning to focus and control attention purposefully. Diamond's studies were conducted in more than twenty classrooms in a low-income school district, and they compared the effects of Tools of the Mind on EF competencies with the effects of the standard balanced literacy curriculum of the school district, which covered similar academic content but did not address EF development. To rule out possible differences in teacher qualities, all classrooms received identical resources and were staffed by teachers with equal amounts of training and support. Likewise, all the children came from the same neighborhood, were randomly assigned to the two programs, and were similar in age and background.

In their second year of preschool, when children in the two programs were compared on the standard cognitive and neural tests of EF, the Tools curriculum was the winner by a substantial margin. And it was most effective for the children who had begun with the lowest levels of EF. Indeed, the children's progress in the Tools program was so impressive that after the first year the educators in one of the schools insisted on ending the experiment so that the children in the control group, who had been receiving the standard balanced literacy curriculum, could also participate in the Tools program.

The opportunity to influence EF development through interventions is not limited to the preschool years. At age 11 to 12 years, with just a few hours of training, children who were underperforming in school were helped to use specific *If-Then* implementation plans and strategies to significantly improve their schoolwork, grade point average, attendance, and conduct. In another study, children with ADHD went through five weeks of training to improve their "working memory" — the memory needed to briefly retain information, like a seven-digit phone number you hear and are trying to keep in mind long enough to dial. Working memory is a crucial component of EF that is required for goal pursuit. This training not only improved their working memory but also reduced their ADHD symptoms and problematic behaviors.

Simple meditation and mindfulness exercises can also substantially improve executive function. "Mindfulness training" helps individuals achieve a present-centered concentrated attention: you let yourself effortlessly become aware of each feeling, sensation, or thought that arises, accepting and acknowledging whatever you experience, nonjudgmentally and without elaboration. In one group of young adults who had five days of training for about 20 minutes a day, these exercises, along with brief meditation, decreased negative affect, eased fatigue, and reduced psychological and physiological responses to stress when compared with a control group that spent the same amount of time doing standard relaxation training. Mindfulness training also reduced distracting thoughts, facilitated concentration, and improved performance in college students on standardized tests like the Graduate Record Examination, used by many graduate schools in the United States as an admission requirement.

Likewise, the normal adult and aging brain can benefit from relatively simple interventions to enhance EF. Two of the most notable are physical exercise, even in moderate amounts and over short time periods, and virtually anything that minimizes loneliness, provides social support, and strengthens the individual's ties and connectedness to other people.

IMPLICATIONS: A SCIENCE CONSENSUS ON PUBLIC POLICY

In short, it has become clear that effective interventions are available to enhance EF. According to the National Scientific Council on the Developing Child, the public policy implications are equally clear. This council consists of a group of respected scientists who have been studying the toxic effects of chronic stress, which typically characterizes the lives of children who live in extreme poverty. They have also examined the interventions that have the potential to substantially reduce those stress levels. In 2011, they reached an unambiguous consensus: strong executive function is crucial for children to build lives that let them develop to their full potential. In light of rapidly growing and convincing evidence "that these capacities can be improved through focused early intervention programs, efforts to support the development of these skills deserve much greater attention in the design of early care and education programs."

As far as science board recommendations go, this message is as passionate and urgent a call to action as it gets. Their conclusions stick closely to the data from the best research and avoid any trace of emotion — which may be why their recommendations too often remain mostly buried in the research archives, with nods of approval from other researchers, or in occasional opinion pieces in the media that cheer them on. The editorials add the passion and duly note that the vast achievement gap in our society is heartbreaking and life destroying to those who live at the bottom and are the "subjects" of this research. There are countless preschool-age kids who do not know the difference between a book's front cover and its back cover, as one pundit put it, who live without ever being told stories or having their imaginations stirred, who hear and have few conversations, who walk hungry to poor schools through dangerous streets, and who return to homes filled with the blaring TV or broken family fighting. High stress is chronic in these children.

Seeing the magnitude of this reality, caring innovators are working hard to make the scientists' messages and recommendations come alive. Many are trying to incorporate what has been learned in the research about self-control, resistance to temptation, and brain development into educational curricula. Some of these efforts are shaking up how educators are thinking about their programs as they try to make education for self-discipline and emotional well-being more effective, beginning with preschoolers.

SOCIALIZING COOKIE MONSTER

One of the best-known efforts designed by innovators in early childhood education is *Sesame Street*, the educational preschool series produced by Sesame Workshop. The program airs worldwide and is aimed at educating and entertaining preschoolers. I have recently had the privilege and pleasure of consulting with the outstanding Education and Research group at Sesame Workshop on how to model self-control skills by trying to socialize Cookie Monster. I emphasize *try* to socialize him, because Cookie Monster definitely has a mind of his own. He personifies untamed visceral desire, specifically for cookies, preferably chocolate chip. He is driven by a hot system only loosely connected to a still primitive prefrontal cortex that seems devoted mostly to assisting him in his search for more cookies, with little interest in helping him inhibit his very hot cookie-specific impulses. This wide-eyed, impulsive blue character has an undisciplined, assertive, extraverted personality. Loudly and proudly announcing "Me want cookie! Me eat cookie!," he proceeds to devour any cookie that is within reach. In its forty-third and forty-fourth seasons, *Sesame Street* sets out a challenge for him: to control his unbridled impulses by cooling his hot system so that he can gain entry into the refined and exclusive Cookie Connoisseurs Club. Preschoolers learn lessons from watching him that illustrate how the findings on self-control can inform and guide the contents and educational mission of preschool programs.

In one segment, Cookie Monster appears on-screen as a game show contestant. In the background, an island-style limbo band sings, "Good things come to those who wait." The congenial but firm game show host asks Cookie Monster if he is ready to play the Waiting Game.

> COOKIE MONSTER: Waiting Game?! Oh boy! Imagine me luck! Me get to play Waiting Game! What Waiting Game?
>
> HOST: The game where we give you a cookie! [A cookie on an easel rises into view.]
>
> COOKIE MONSTER: Oh boy! Me love this game. Coo-kie! Ahm! [Cookie Monster rushes to devour it but the host grabs it away.]
>
> HOST: Wait!
>
> COOKIE MONSTER: Wait to eat cookie? That crazy talk! Why me wait?
>
> HOST: Because this is the Waiting Game and if you wait to eat the cookie until I get back, you get two cookies!

The lesson proceeds, as the host patiently explains the rules again: "If you wait for the cookie till I get back, you get two cookies!" For about a second, Cookie Monster thinks it's a good idea: "Me wait then," and the host wishes him good luck, but Cookie Monster quickly has a hot insight — "Oh, who me kidding, me can't wait! Me have cookie now!" — and lunges for it, intercepted by the Waiting Game Singers, who pop up singing, "Good things come to those who wait."

The singers explain that singing is a good strategy to use when it's really hard to wait for something, and Cookie Monster tries, but can't and doesn't want to. "Forget it, me just eat it!" The singers intervene again: "You need another strategy. Remember, good things come to those who wait. Yeah, good things come to those who wait."

The lesson continues as Cookie Monster learns to pretend the cookies are in a frame, traces a mental frame with his fingers, pulls out a real picture frame and frames the cookie with it, twiddles his thumbs, hums *dum dee dum,* but soon becomes tempted again. He keeps getting support as he learns new strategies, step by step, and is amazed as he begins to discover some by himself: "Me need another strategy. Ahh! Got it! Me take me mind off cookie by playing with this toy." He brings up a stuffed dog and begins to sing to himself and play with it, until that gets boring and he invents a new way to keep going: "Me pretend delicious cookie is a very smelly fish," as the cookie is transformed into a fish on the easel and he waits, waving the air around as if smelling a stinky odor. In time, lots of time, after much effort and with increasing grit, he wins the Waiting Game and joins in the music, singing triumphantly, "Good things come to those who wait."

This episode is one of many in the two years of programming that Sesame Workshop is devoting to self-regulation. Its 2013 and 2014 seasons of *Sesame Street* provide memorable, entertaining lessons about the diverse forms of self-regulation conveyed in the antics and adventures of its adorable and long-beloved characters, from Cookie Monster to Oscar the Grouch in his garbage can. They engage preschoolers in funny, short stories while teaching some of the most essential aspects of self-control and a host of other strategies and skills preschoolers need to begin to develop executive function, self-restraint, and the regulation of their own emotions.

The *Sesame Street* education researchers have made many efforts to objectively assess the impact of their programs, and they have gathered evidence over the years documenting their program's connection to many positive outcomes, including greater school readiness and success. Although kids who watch more *Sesame Street* do better, we cannot know if that's because of what the program teaches or because they are children whose parents are more likely to turn the television to educational programming. Most likely, both factors contribute to making these programs useful — and not just for keeping children busy and happy, but for helping them develop skills and learn important social, moral, and cognitive life lessons.

FROM COOKIE MONSTER TO KIPP SCHOOLS

Leading scientists who worry about the effects of toxic stress on the infant's brain and on subsequent susceptibility to mental and physical illness note that those who are lowest in socioeconomic status (SES) have greater morbidity and mortality for diverse diseases and suffer from what has been named the "biology of disadvantage," or the physiological and psychological consequences of living under chronic stress, beginning at conception. For educators who work with people at the very low end of the SES scale, the challenge is how to help children, parents, and caregivers overcome that disadvantage. The most promising route is to provide access to education as early in life as possible, which in turn can help them climb up the SES ladder. But what kind of education and with what methods?

The depressing state of public school education in the

United States, particularly in impoverished school districts, has received widespread attention. The encouraging news in an overall grim picture is that in the past ten years or so, and with accelerating speed, diverse innovative educational interventions are being developed that try to incorporate what has been learned about brain development, delay of gratification, self-control, and self-discipline into curricula. Many of these ventures are working to make education more effective in different types of school settings, particularly those that children with the biology of disadvantage generally attend.

Here I focus on one promising effort that closely connects what it teaches to findings at the cutting edge of psychological science: the KIPP school programs in New York City, which helped George Ramirez find his way. In the fall of 2012, I visited four of the nine KIPP academy schools then in New York City, with a tenth under construction. That KIPP stands for the Knowledge Is Power Program is proudly announced on signs throughout its schools. I went to glimpse how the program was faring in the real world as it tried to educate children living in some of the poorest SES areas of the country. My goal was to get a sense of what was possible in this kind of school.

KIPP is becoming a model for different kinds of efforts to transform public education. My introduction came from Dave Levin, the seemingly inexhaustible forty-something engine that's driving the KIPP group of charter schools. These schools are devoted to preparing children, starting in kindergarten, for college, and college banners hang all over their classroom walls. More than 86 percent of the students are inner-city minority children from impoverished backgrounds. They arrive at 7:30 a.m. and are dismissed at 4:30 or 5 p.m. In the summer there are two to three weeks of additional school. Parent participation and parental visits are encouraged in many programs. The children are selected by lottery, since there are not nearly enough places for the many who want and deserve a chance. The New York City KIPP schools are modeled on a program that Dave Levin and Mike Feinberg started in one fifth-grade classroom in Houston, Texas, in 1994. In 2014, there will be 141 KIPP schools nationwide with about fifty thousand K–12 students.

One of the schools I visited, KIPP Infinity Elementary School, is located in the predominantly Hispanic and African American neighborhood of Harlem in Manhattan, a few blocks north of Columbia University and south of the City College of New York. This KIPP school opened in 2010 and has about three hundred students, from kindergarten to fourth grade, of whom more than 90 percent are African American or Hispanic, and about the same percent qualify for the free or reduced-cost lunch program available to low-income families. The school is exceptionally attractive, sparkling clean and well lit, with comfortable, modern furnishings and equipment. Having attended New York City public schools when I was a child and visited them for research in recent years, I found the contrast in appearance alone a happy surprise.

When I wandered into a first-grade classroom I saw kids attentively listening as their young teacher spoke quietly with them. I was approached and greeted immediately by "Malcolm," a little boy with a soft voice and gentle manner who politely introduced himself. While extending his hand for a warm handshake, Malcolm asked my name and welcomed me to the classroom, the Columbia University Lions. As he ushered me in, loud drumrolls and lots of cheering erupted as the teacher announced who had been selected for Name Day that morning — not because of a birthday, but a sweet, enthusiastic celebration of a different child every day.

Each classroom is named for a different college and has banners with inspirational themes hanging on the walls that are discussed repeatedly. UNITE, for example, is the acronym for Understand, Never give up, Imagine, Take a risk, Explore. A "recovery chair" or "thinking chair" sits in one area of the room, not for the standing-in-the-corner punishment of earlier times but to help students cool down when they feel that they're about to lose it or for when the teacher believes that is about to happen. The area around the chair includes a timer with sand flowing through and messages displayed on the nearby wall to help the child self-soothe: get distance from a hot situation, breathe deeply, count backward, imagine anger floating away in helium balloons, and other strategies for calming down, regaining control, and going from feeling hot to thinking cool so that she can leave the chair and return to rejoin the class.

"Madeline," age ten, was in the fifth grade, nearing the end of her first year at KIPP, when I met her. She had moved to KIPP from the public school on the other side of the building. "Over there it was more cold," Madeline said about the public school, "and here the teachers are more strict, with more expectations that you have to follow." Her enthusiasm kept spilling out: "I think I'm learning differently — the teachers are clearer. Each day we learn new and review the old. Here we take school more seriously. More homework, review more, get reports on how we are doing. Progress reports — you still have a chance of changing it if you attend better, behave better. A report card is just a final grade."

What will she be doing when she's 20 years old? She'll be a doctor or a vet or a teacher, she said. How will she get there? She answered thoughtfully, slowly, with many details and examples, from "The more I listen, the more I learn," to spending three hours on homework each night, to reflecting on herself and how she was changing: "I'm learning more, becoming a more hard-working person.... We have 90 minutes in each class and learn something new every day."

"What's social intelligence?" I asked her. Her answer: "Like when something falls and you pick it up before you're told. It's when you think ahead before someone tells you. If someone is behaving badly in class you don't listen to him." What is self-control? "It's similar to social intelligence. Even though someone is doing something that's funny in class, you don't laugh — you have to control yourself. If you want to take something you want, you control yourself and don't." She reminded me of the answer I got from another child the same age who was struggling to gain better self-control: "It's thinking before doing," he had patiently explained.

As a researcher, I know that I can't generalize from a small sample; I realize that I have to be cautious not to reach hasty conclusions from very thin slices of behavior, and I must temper my impressions with caveats. But I also know that wandering through KIPP classrooms, meeting these children, and catching glimpses of how they listen and speak, as well as how the teachers teach, left me feeling a lot more optimistic about the future for at-risk children.

I felt more than a warm glow. My cool system saw that, when

wisely applied by dedicated teachers in the right classroom environment, the lessons learned in the lab could give these children the chance to change their lives, discover their goals, and work hard to progress toward them. KIPP exemplifies an educational philosophy and school system that is incorporating research findings into its daily curriculum and way of life. It is demonstrating that self-control can be nurtured, goal setting encouraged, realistic goals achieved, curiosity stimulated, and persistence rewarded until grit becomes its own reward.

I asked Dave Levin if KIPP schools really "save lives," to use George Ramirez's phrase. Dave was adamant that they *don't* save anybody's life. He insisted: "We're the cheerleaders; the kids are playing the game. They do the heavy lifting. We set up the conditions; the hard work has to be done by each individual." KIPP's mission, he explained, is to help children have choice-filled lives. Choice does not mean one road for all — and it does not have to mean an Ivy League college, or even college at all. Choice is about children having genuine options in how they make their lives, regardless of their demographics.

BUILDING "CHARACTER SKILLS"

Dave and I talk often about how he thinks KIPP needs to evolve to be even more effective and how it is changing. In the 1990s, when KIPP began, college and the academic training required to get there seemed to be *the* passport out of toxic poverty and into a world of opportunities and choices. Therefore, KIPP's overarching goal was, and remains, to do whatever it takes to get its students to complete college. Dave tells me that in 2013, about 3,200 KIPP graduates were in college, with a cumulative college completion rate of about 40 percent. This compares with a rate of 8 to 10 percent for children of similar backgrounds who are not in KIPP programs, and a United States national average college graduation rate of 32 percent.

Dave believes that this success rate reflects the fact that KIPP students not only learn academic skills required for college but are also taught the character skills necessary to thrive there and beyond. For him, the continuing challenge is how to most effectively build "character education" into the KIPP curriculum. I was worried when he first mentioned "character" because so often the term is used for inborn traits, but that is not what it means in these schools. Instead, character is viewed as a set of teachable skills, specific behaviors and attitudes — most important self-control, but also such qualities as grit, optimism, curiosity, and zest. KIPP schools are trying to make character education more than just posting inspirational slogans throughout classrooms and having school principals deliver lofty weekly sermons during assembly; rather, they are trying to make it an integral part of the daily learning experience of all their students, and equally of their teachers and mentors.

I asked Dave how KIPP manages to make character education tangible, to give it teeth, in the classroom. He believes that the key is to give students the opportunity to practice in school the critical behaviors that nurture self-control, grit, and other character skills. In his words: "If you want kids to learn how to get over frustrations quickly, bounce back from failures, and work independently with focus, they have to be given the chance to do these things in their academic classes, and the teachers need to structure their lessons to allow time for this." Therefore, the curriculum provides substantial time for practice in which the students work independently, with a partner, or in a small team — but independent of the teacher — on challenging projects that demand concentration and sustained effort. "The key is that the teacher is no longer standing in front of the class talking but rather forcing kids to do heavy lifting."

To monitor their progress in character education, students evaluate themselves several times a year, at the end of each marking period. They evaluate how often (from "almost never" to "almost always") they successfully practiced a set of behaviors that define each character skill — specifically self-control, grit, optimism, zest, social intelligence, curiosity, and gratitude. Each skill is linked to the defining behaviors with phrases like "I stayed motivated, even when things didn't go well" for optimism and "I finished whatever I began" for grit. Self-control is divided into two types of self-discipline: the ability to keep goals in mind and stay focused when working ("I paid attention and resisted distractions") and the ability to control temper and frustration in upsetting interpersonal situations ("I remained calm even when criticized or otherwise provoked"). For "zest" the behavioral features are items like "I approached new situations with excitement and energy." And social intelligence is defined with behaviors like "I demonstrated respect for the feelings of others." The teachers are asked to observe and rate the progress not only of their students but also of themselves on similar measures of character development to assess the progress of the entire school community and guard against decline. These efforts to enhance character skills have not yet been systematically evaluated, but the children and teachers in these programs are at least beginning to think and talk in this language about whether they are successfully building the desired character skills.

As I learned about the character skills that KIPP is working hard to develop in its students, I was struck by their similarity to the qualities that differentiated the preschool children who were able to wait on the Marshmallow Test from those who quickly rang the bell when we looked at them as adolescents a decade later (as discussed in Chapter 1). Take "grit," for example, which is measured on Angela Duckworth's "grit scale" with items like "Setbacks don't discourage me." This statement is almost verbatim one of the qualities that parents said characterized their adolescent children who waited longer on the Marshmallow Test in preschool. It is encouraging to see the overlap in the behaviors and attitudes that distinguished high-delay children as they grew up and the character skills that KIPP is trying to enhance in its students to improve their chance for a successful future.

For many reasons, schools like KIPP often start in kindergarten and not in the earlier preschool years, when children are highly vulnerable and the biology of disadvantage lays down its roots. Preschool is also the time when children are most ready to learn strategies that can help them cope with stress and develop cognitive skills essential for school success. To help narrow the ever-growing achievement gap between the well-off and the poor, President Obama called for making preschool education universally available in the United States in his 2013 State of the Union address. If this call to action translates into reality, the success of the effort will depend in part on how effectively preschools incorporate lessons from the research into their work. And while preschools can help provide the essential foundations, long-term gains will depend on how schools and families collaborate to help children continue to use and further develop skills to generate the conscientious behaviors, self-control, responsibility, and life goals that society values. It remains to be seen how the Obama proposal for universal access to preschool in the United States will fare. But there is good reason to believe that better access to preschool, however it can be accomplished, is urgently needed, and to hope and advocate that if it materializes, these schools will help young children develop the character skills and motivation they need to have the chances they deserve.

19

APPLYING CORE STRATEGIES

THE SELF-CONTROL CONCEPTS AND strategies in this chapter will not be news to you since I discussed research about each of them throughout this book. In this chapter, I put them together, show how they connect, summarize key points, and focus explicitly on how they can be applied in everyday life to help with self-control efforts, when and if you want to try them out.

To begin with, resisting temptation is difficult because the hot system is heavily biased toward the present: it takes full account of immediate rewards but discounts rewards that are delayed. Psychologists have demonstrated this "future discounting" in both humans and animals, and economists have formalized it in a simple mathematical model. David Laibson, a Harvard University economics professor and my colleague in ongoing research, has used it to explain why he rarely makes it to the gym, in spite of his good intentions to go there regularly. Individuals differ in how severely they discount the future, and his example uses a discount rate that cuts the value of delayed rewards in half. For most people the discount is even larger. In order to model the discounting, Laibson assigns a number value to each activity, rating how much pain or effort (negative number) or how much reward (positive number) the activity provides. For him, the effort of exercising today has a cost of –6, and the long-term health gains from exercise have a positive value of +8. Of course these numbers always depend on the values of the individual making the decision.

This is how Laibson explains his procrastination: He can exercise today (his effort cost is –6) to gain delayed health benefits (for him a future value of +8). The net benefit of exercising today for someone with his present bias is (–6 + ½ [8] = –2). In this equation, the future value of +8 was halved because of the automatic discounting of the future, making –2 the net benefit of his exercising today. In contrast, exercising tomorrow has a delayed effort cost of –6 and a delayed benefit of +8, both of which are halved because they are in the future (½ [–6 + 8] = +1). For Laibson the resulting net value for putting off going to the gym is +1, which is better than the –2 net value for exercising today. Consequently, he is rarely at the gym. These weightings vary a great deal not only among individuals but also within each of us for different activities: you may religiously make it to the gym but always avoid cleaning your closet.

The emotional brain's predisposition to overvalue immediate rewards and to greatly discount the value of delayed rewards points to what we need to do if we want to take control: we have to reverse the process by cooling the present and heating the future. The successful preschoolers demonstrated how to do this. They cooled their immediate temptation by physically distancing themselves from it. They pushed it to the outer edge of the table, turned around in their chairs to face the other direction, and invented imaginative ways to purposefully distract themselves, all while keeping their goal (two marshmallows) in mind. In experiments in which we suggested cooling strategies to help them delay for the larger rewards, they cooled the immediate temptation by transforming it cognitively, rendering it more abstract and psychologically distant, which made self-control much easier for them and let them wait longer than we could even bear to watch.

THE FUNDAMENTAL PRINCIPLE: COOL THE "NOW"; HEAT THE "LATER"

Regardless of age, the core strategy for self-control is to cool the "now" and heat the "later" — push the temptation in front of you far away in space and time, and bring the distant consequences closer in your mind. My colleagues and I demonstrated this in the experiments with tobacco and food cravings described in Chapter 10. When we cued participants to focus on "later" and the long-term consequences of eating ("I may get too fat"), they experienced reduced food cravings, both in what they felt and in what their brains registered. Likewise, when heavy smokers focused on "later" and the long-term consequences of cigarette smoking ("I may get lung cancer"), their tobacco cravings diminished. Focusing on "now" and the immediate, short-term effect ("It will feel good") of course had the opposite effect, making the cravings impossible to resist.

Outside the lab, when our hot system makes us focus on the present temptation in life, there is no one to cue us to make the distant consequences hot and the immediate gratifications cool. To master self-control, we have to instruct ourselves. And that won't happen naturally because in the face of temptations, the hot system dominates: it discounts delayed consequences, it activates faster than the cool system, and as it accelerates the cool system attenuates. This dominance of the hot system might have served our ancestors well in the wild, but it also drives us to the default reflex of giving in to temptations, making it easy for smart people to behave stupidly. If we feel regret about our self-control failures, it will probably be fleeting, because our psychological immune system is so good at protecting and defending us, rationalizing our lack of self-control ("I had a crazy day"; "It was her fault"), and not letting us feel bad about ourselves for long. That makes it even more unlikely that we will learn to behave differently in the future.

IF-THEN IMPLEMENTATION PLANS MAKE SELF-CONTROL AUTOMATIC

How do we get around this problem? If we want to exert self-control we have to find ways to activate the cool system automatically when we need it, which is exactly when this is hardest to do unless we have prepared for it. Recall how the young kids resisted the seductive Mr. Clown Box, who kept urging them to come talk and play with him now rather than continue their work and play later (Chapter 5). They prepared for the encounter by first rehearsing *If-Then* implementation plans. For example: "*If* Mr. Clown Box makes that *bzzt* sound and asks you to look at him and play with him, *then* you can just look at your work, not him, and say, 'No, I can't; I'm working.'" Such *If-Then* plans helped the children stick to their goals, persist in their work, and resist Mr. Clown Box's beguiling temptations.

In life, employing *If-Then* implementation plans has helped adults and children control their own behavior more successfully than they had imagined possible. If we have these well-rehearsed plans in place, the self-control response will become automatically triggered by the stimulus to which it is connected ("*If* I approach the fridge, *then* I will not open the door"; "*If* I see a bar, *then* I will cross to the other side of the street"; "*If* my alarm goes off at 7 a.m., *then* I will go to the gym"). The more often we rehearse and practice implementation plans, the more automatic they become, taking the effort out of effortful control.

FIND THE *IF* FOR *IF-THEN* PLANS

The first step in creating an *If-Then* plan is to identify the hot spots that trigger the impulsive reactions you want to control. In the Wediko camp studies (Chapter 15), the researchers looked not just at how much aggression the children expressed but also at the psychological situations in which they did and did not express it. Rather than being broadly consistent across many different kinds of situations, problematic behaviors were highly contextualized and depended on the specific type of situation. While "Anthony" and "Jimmy," for example, had similar overall average levels of aggression, the hot spots that triggered each of their outbursts were completely different. Anthony was explosive *If* interacting with peers even if they were nice to him, while Jimmy lost control *If* interacting with adults, but not with his peers, even if they teased and provoked him.

One way to identify our own hot spots is to keep a journal to track moments when we've lost control, similar to the self-monitoring I described in Chapter 15 to track stress reactions. People tracked the specific psychological events that triggered their daily stresses, identifying each situation in which their stress occurred and noting its intensity. Their hot spots were usually more specific than they had expected them to be. Recall that "Jenny," for example, discovered that she had stress levels no higher than average in most situations, and often they were below average; her stress levels were only extremely high in situations in which she felt excluded. That is when she fell apart, full of anger at others as well as at herself. Once we identify our precise hot spots, for example by tracking them as they are triggered, we can begin to form and practice specific *If-Then* implementation plans to change how we cope with them.

For rejection-sensitive "Bill" (featured in Chapter 12), an especially troubling situation that triggered his anger was when he felt his wife focused on the newspaper rather than on him at breakfast. He could practice implementation plans for this scenario so that when she turned to the headlines, he automatically activated a cooling strategy to self-distract, such as silently counting down from 100 until he had calmed enough to inhibit the destructive outburst he was about to have. He could then substitute a constructive alternative ("Please pass me the business section"), thereby helping himself, step by step, to maintain the relationship he was afraid to lose. It sounds simplistic, but it can be astonishingly effective in actual practice, as Peter Gollwitzer and Gabriele Oettingen have repeatedly documented in their research. The hard part is maintaining the change over time, which is true for most efforts to enhance self-control, from dieting to giving up cigarettes. If we persist, however, the gratification that our new behavior produces will help sustain it: the new behavior itself becomes valued, no longer a burden but a source of satisfaction and self-confidence. As with all efforts to change long-standing patterns and learn new ones, whether playing the piano or exercising self-restraint to avoid hurting the people we love, the prescription is to "practice, practice, practice" until it becomes automatic and intrinsically rewarding.

PLANS SURE TO FAIL

When people anticipate that they will not be able to control themselves, they often try to make precommitments to reduce the temptations in their environment: they strip the house of the irresistible foods that are bad for them, get rid of the liquor, or throw away their stashes of cigarettes, and resolve not to buy more of the tempting products — or, if they do buy them, they do so in smaller, more expensive quantities in the hope that this will make them too costly to afford. Precommitment strategies — from Christmas savings clubs to insurance policies and pension plans — can be relatively low-cost ways to reap valuable benefits. But when these strategies are tried without a binding commitment, without a specific *If-Then* implementation plan to give them teeth, they are likely to go the way of New Year's resolutions. We are wonderfully creative at making tepid commitments and then finding endless ways to get around them.

I saw this play out with a friend and colleague, now long deceased. He was a renowned research psychologist who was half-heartedly trying to cut down his incessant smoking by using the precommitment strategy of refusing to buy any cigarettes. Instead, he mooched them from anyone nearby. At Christmastime, the offices at Columbia University were mostly empty, and this restricted his efforts; but in his desperation, he began searching for cigarette butts on the sidewalks of Manhattan. He described to me his greatest moment of shame: He had at last detected a cigarette butt on Broadway that looked tempting and he stooped down to get it. As he rose with the butt in his hand, he saw the look on the face of the street person who was always on that corner. The street person had been reaching to pick up the same butt but was not fast enough, and now exclaimed to my elegantly attired friend, "I don't [expletive] believe it!"

My friend illustrated how to make precommitments that are guaranteed to fail, even though he was more than smart enough to understand this himself. Instead of enlisting other people to help him stick to his ostensible goal, telling them to refuse to give him cigarettes no matter how much he begged, their polite responses to his mooching helped him fail. He was fully aware that in order to fight the power of the immediate temptation to smoke, he would have to make the costs of violating his precommitment to quit much greater than the value of getting his cigarette immediately when his hot system wanted it (which was most of the time). As psychotherapists, whatever their orientation and strategies, regularly tell their clients, you have to want to change, with emphasis on the "*want* to."

PRECOMMITMENT PLANS THAT CAN WORK

To make precommitment strategies work, turn them into *If-Then* implementation plans. Many examples can be found in cognitive behavior therapy. In my friend's situation, he would have had to precommit by leaving huge checks made out to his most hated causes (of which he had many), with a binding contract authorizing his therapist to mail one check each time a cigarette was mooched or smoked. If this is a strategy you want to try without a therapist, you can ask an accountant, lawyer, closest enemy, or best friend to mail your checks.

The steep discounting of future rewards takes a tragic toll on everything from health care to retirement planning. Millions of people in the United States, for example, are shocked at how little they have saved when their distant future self becomes their present self at age 65 (as discussed in Chapter 9). Recognizing the scope and seriousness of this problem, researchers have helped employers bypass the limitations of human self-control by making retirement savings the default option when new employees sign on to their firms. At one large firm, the rate of participation in the 401(k) retirement plan after one year of service was 40 percent when nonenrollment was the default; when the default was enrollment or required action to opt out, it was 90 percent.

If we don't have such forward-looking employers, we can try in our halcyon days to connect more closely with our future selves, to keep who and what we are trying to become in mind, to construct a life story that has continuity and direction and long-term goals that become visible not just when looking back but also when looking ahead. At the concrete action level, we can use implementation plans to nudge ourselves and select the option to save the highest percent we can afford for the retirement plan, to be initiated the day we are eligible to sign on at a new job. Or, if we are still at the old job, we can implement a plan to check in with human resources on Monday morning at 10 a.m. to be sure that the right retirement savings option is in place and saving for us automatically. Such strategies can help us get around the discounting equation — assuming that the retirement plan does not default by the time we need it, and that we are still around to use it.

COGNITIVE REAPPRAISAL: IT'S NOT A TREAT; IT'S POISON!

In Part I, we saw that how the preschoolers mentally represented the temptations determined how well they could control themselves. When they reappraised their hot temptations to cool them down, they were able to wait for their delayed treats. Twenty years later, I had a "Eureka!" moment in which I suddenly realized the meaning of those findings in my own life. This was in 1985, but it is still as vivid as if it were yesterday.

A terrible itchy rash had erupted around both elbows, as painful as if I had dipped them in acid. It kept spreading and getting worse, and after a year of agony I found a well-known dermatologist who explained that my problem was a version of celiac disease. The prescription he gave me would help manage it, and I would have to take it indefinitely. He cautioned that frequent blood tests were needed to monitor against the possibly serious side effects of the medication. I soon improved, but a milder version of the rash persisted. After many months, I learned from the medical school library (this was before Google and easy Internet research) that, while little was known about celiac disease at the time, it was due to an autoimmune reaction to gluten, contained in wheat, barley, and rye. A gluten-free diet was the only cure; although the drug my physician gave me could alleviate the symptoms, it would not prevent the long-term destructive effects of the disease.

I asked my dermatologist why he had not told me that I should be on a gluten-free diet. Nobody has the self-control needed to stay on a gluten-free diet in a gluten-filled world, he said, so there was no point in talking about it. A quarter of a century later, it turns out that a great many people all over the world are being diagnosed with celiac disease — and lots of them, myself included, are managing to stay on gluten-free diets. People are able to do this not because they are exceptionally good at self-control but because the news that gluten is toxic for them has shifted the reward values in their discounting equation. Previously irresistible temptations, like chocolate fudge cakes, French baguettes, and pasta Alfredo, have suddenly become their poison.

The fact that the consequences of merely tasting anything with gluten usually come quickly, surely, and painfully to those with celiac disease, of course, makes the transformation much easier — almost automatic. For behaviors like forgoing tobacco, dieting, controlling one's temper, or saving one's income for retirement rather than spending it, the negative consequences are in the distant future and probabilistic, rather than swift and certain. They are abstract, unlike a painful burning itch or gastrointestinal distress. Therefore, you have to reappraise them to make them concrete (visualize your lungs with cancer on an X-ray the doctor is showing you as he gives you the bad news) and imagine the future as if it were the present.

SELF-DISTANCING: STEPPING OUT OF THE SELF

The best self-control plans notwithstanding, anger, anxiety, rejection pain, and other negative emotions are inevitable parts of life. Consider the heartbreak of people who are rejected by a partner or spouse after years of building a committed life together (Chapter 11). Many people who have been hurt this way keep reliving their terrible experiences, refueling their sadness, anger, and resentment, and depressing themselves more deeply. As their stress increases, the hot system becomes even more dominant, deactivating the cool system and triggering a vicious cycle: increased stress → hot system dominance → negative emotions → long-term distress → deepening depression → loss of control → chronic stress → increasingly toxic psychological and biological consequences → increased stress.

To get out of this trap, it can help to temporarily suspend the habitual self-immersed view we have of ourselves and the world. You look again at the painful experience — not through your own eyes, but as if you were observing from a distance, like a fly on the wall, observing what happened to a third party. This change in perspective alters how the experience is appraised and understood. By increasing your psychological distance from the event, you reduce stress, cool the hot system, and can use the prefrontal cortex to reappraise what happened so that you can make sense of it, gain closure, and move on.

The mechanisms that enable these changes are still being studied, but the shift from self-immersion to self-distancing significantly reduces psychological and biological distress and lets us regain better control of our thoughts and feelings. That makes the mental acrobatics of becoming the observant fly on the wall worth a try. It is not easy to do by yourself, but cognitive behavior therapy uses many of the principles and findings discussed throughout this book to help with the toughest struggles. This kind of therapy can be especially useful when one's own self-control efforts fail, as they did in John Cheever's story "The Angel of the Bridge." When the hot system has formed intense anxiety-producing associations that are triggered automatically, it generates disabling panic. Without help, these negative associations can resist even the best self-control efforts, unless we are lucky enough to meet our own angel of the bridge.

WHAT CAN PARENTS DO?

At the end of every talk I give in schools, after I emphasize that self-control is far from completely prewired, parents ask, "What can we do to help our children?" When there is enough time, I begin by telling them that it is especially important to keep stress levels low during pregnancy and in the infant's first few years. It is well-known that exposure to extreme and prolonged stress early in life can be terribly damaging. More surprising is that children who live with seemingly mild chronic stressors in their first year of life, like exposure to persistent, albeit nonviolent, parental conflict, may experience increased stress reactions in their brain simply when they hear angry voices while they are sleeping. To keep infants' stress levels low, a first step for parents might be to try to reduce their own stress, recognizing that it often increases when newborns arrive. The same strategies that cool and control hot system reactions to impulses, temptations, and rejection experiences apply when dealing with crying and needy infants every few hours in the middle of the night, especially when you are exhausted yourself.

Beginning in the first year, caretakers can use distraction strategies to turn the child's attention away from feelings of distress and focus instead on diverting stimuli and activities. In time, the young child learns to control her attention to reduce her own distress by self-distracting, a basic step for developing executive function. Parents can be helpful guides in this transition. "Bruce," a writer who worked at home, spent much of his time caring for his four-year-old son. Once, when the boy was waiting for his favorite television program and it refused to come on-screen when he wanted it to, he dissolved in a temper tantrum. Bruce had heard about the marshmallow research and how self-distraction helps kids wait for their treats, so he decided to try it with his son. He calmed his son down and suggested that there were ways to make waiting much easier: just distract yourself and do other fun things in your head or for real until the program comes on. His son then picked up his favorite toys, moved away from the television set, and played happily until his program appeared. Bruce was surprised at how easy it was and delighted when he saw that his son seemed to have learned from this experience, as he continued to use self-distraction to make delay more manageable in other situations.

Distraction won't work when young kids are hurting one another, especially when the caretakers are not close by. "Elizabeth" is a licensed, practicing therapist and counselor who is well trained in cognitive behavior therapy and often works with children who are having self-control struggles, and with their parents. I asked her for examples of strategies she uses to help preschoolers control aggressive behavior, and she gave me some tips from her efforts to help her own son, who was three at the time. She told me,

> He was biting, sometimes up to three kids a day at school. After trying many strategies, what finally worked was a very simple "Boys who bite do not get dessert" policy. So I'd pick him up at school, find out if he had bitten; if so, then no dessert that night. We talked about it ahead of time, rehearsed it again on the way to school, and on the first day he heard me tell his teachers the policy before I left. When I got to school to pick him up, he had had one biting incident, near the very end of the day. I told him, "OK, no dessert tonight." He said, "OK, Mama," and we hugged. When we got home, I showed him the dessert I had made. I reminded him that if he did no biting the next day, he could have some that evening. He understood. Each time there was an incident we'd brainstorm alternative strategies and think of things he could do instead of biting. We'd practice on the way to school. Whenever he used an alternative strategy ("I used my words, Mama!"), I'd praise him for making good choices. In three or four days he was bite-free, and since then there have been no further incidents."

Elizabeth's example highlights the importance of helping children learn early on that they have choices, and that each choice has consequences. It also illustrates that rewards can be used judiciously to encourage appropriate choices. What those rewards should be depends on the parents' values and what will work for the particular child. Parents who want to avoid using food as rewards, for example, can easily find other treats and experiences.

The self-control strategies that children develop are shaped by their attachment experiences with caretakers from the start of life. Parents who hope to raise children who will stay closely attached to them but also develop adaptive self-control skills can improve their chances through their own behavior. If they are sensitive to their toddler's needs, provide support and help when it's wanted, but also encourage autonomy, their odds will be better than those of parents who intrusively overcontrol their children or are more focused on their own needs than on their children's (Chapter 4).

To promote children's sense of both autonomy and responsibility, we can help them realize early in life that they do have choices that are theirs to make, and that each choice comes with consequences: good choices → good consequences; bad choices → bad consequences. Recall George Ramirez, who felt lost and adrift in his chaotic life as a young child in the South Bronx and then went on to become a successful student at Yale University. He dated the beginning of his own "saved life" to the time at age nine when he learned his first lesson about the causal connection between his choices and their consequences. On his first day at KIPP, George began to see that he actually had choices, that it was up to him to make them, and that it was his responsibility to deal with their consequences. It was the teachers' responsibility to make sure that his choices resulted in the consequences they deserved. It was the same *If-Then* lesson that Elizabeth taught her young son about biting other kids: children who bite don't get dessert. George's lesson was that third graders who don't listen don't learn, and "if I'm polite to others, they're polite to me" (Chapter 8).

Parents can do much to create conditions in which their young children succeed. One important strategy involves working with them on enjoyable but challenging tasks that become increasingly difficult, whether it's learning to play the piano, building with blocks and Legos, or climbing on the jungle gym. The challenge for the parents is to provide the support their child needs and wants, and then let her work on her own, without taking over and doing it for her. Early success experiences help young children develop optimistic, reality-based expectations for success and competence and prepare them to discover for themselves the kinds of activities that ultimately become intrinsically gratifying for them (Chapter 8).

We can also help children develop "incremental growth" mind-sets in which they think of their talents, abilities, intelligence, and social behavior not as reflecting fixed inborn traits but as skills and competencies that they can cultivate if they invest the effort. Rather than looking for good grades and applauding kids for being "so smart," we can praise them for trying as hard as they can. As Carol Dweck's research (discussed in Chapter 8) illustrated, guiding children to think about their abilities and intelligence as malleable skills prepares them to use effort to improve their performance. Just as important, we can help them understand and accept that failures along the route are part of life and learning, and then encourage them to find constructive ways to deal with such setbacks so that they keep trying instead of becoming anxious, depressed, and avoidant. And if we want them to be willing to delay gratification when we promise them delayed rewards, we'd better be careful to keep our promises.

But arguably the best answer to the "What can we do to help our children?" question is to model what you would like them to become. How parents and other important figures in a child's life do or do not control themselves — how they deal with stress, frustrations, and emotions; the standards they use in evaluating their own achievements; their empathy and sensitivity to other people's feelings; their attitudes, goals, and values; their disciplinary strategies; their lack of discipline — all profoundly influence the child. Parents model and teach children an enormous repertoire of possible reactions to endless challenges, from which children select and transform what fits and works uniquely for them over the course of their own development.

Much research has shown the powerful effects of models, even in short-term experiments, on everything from navigating aggressive feelings in preschool to overcoming a fear of dogs to recovering from heart surgery to avoiding unsafe sex. In preschool experiments, for example, when friendly adult models scolded and beat up a Bobo doll at Stanford University's Bing Nursery School, the preschoolers who had watched them imitated their aggressive behavior in exquisite detail later, when they were left to play by themselves, and even added their own elaborations. Likewise, when models rewarded their own performance in a bowling game only when their scores were very high, and cued the children they played with to do likewise, they strongly affected the self-reward patterns and achievement standards that the kids adopted for themselves later, when bowling by themselves in the models' absence.

Stories about what happens to fictional children, beloved animals like baby bears and cuddly tigers, and animated locomotives who do all kinds of constructive and destructive things that lead to different consequences teach young children lessons about good and bad behavior that they love to hear over and over, again and again. Preschoolers don't know that these bedtime stories and televised educational programs are teaching them executive function. Different characters enact themed stories imparting positive social and emotional values, including how to cope with sadness, how to use words instead of actions to deal with anger, how to be a good friend, how to express gratitude, and how to delay gratification. These books and programs can help young children learn to deal with stress and interpersonal conflicts and develop executive function, all through a medium they enjoy.

Regardless of how the strategies are learned, children are fortunate if by ages four and five they know and use methods that make it increasingly easy and automatic for them to cool their hot systems when they need to — whether it is by playing happily by themselves or by waiting for the bigger treats on the Marshmallow Test. But I can't end this discussion without reiterating: a life lived with too much delay of gratification can be as sad as one without enough of it. The biggest challenge for all of us — not just for the child — may be to figure out when to wait for more marshmallows and when to ring the bell and enjoy them. But unless we learn to develop the ability to wait, we don't have that choice.

20

HUMAN NATURE

"*Your future in a Marshmallow.*" When I first saw this header about my research posted on the Internet, it was the stimulus that prompted me to start writing this book. I Googled the phrase as I began this final chapter, and it led me to "Fate may not be written in the stars, but what if it's written in our genes?" The research in this book tells a story that leads to very different conclusions than this tagline suggests. It is the story of how self-control can be nurtured in children and adults, so that the prefrontal cortex can be used deliberately to activate the cool system and regulate the hot system. The skills that enable this give us the freedom to escape from stimulus control to achieve self-control, thereby giving us real choice — instead of being pushed by the immediate impulses and pressures of the moment. A main lesson from modern science is that rather than being predestined by DNA and development in the uterus, the architecture of our brains is more malleable than had been imagined, and we can have an active hand in shaping our fates by how we live our lives.

While many, indeed most preschoolers who managed to delay gratification on the Marshmallow Test continued to show good self-control for decades to come, self-control steadily decreased for some, whereas others who rang the bell soon showed the opposite pattern over the years, increasing their self-regulation as they matured. This book tries to make sense of that variability, to convey its complexity, and points to some of the choices in the course of development that influence how life unfolds.

EF AND BURNING GOALS

The children who persisted in the Marshmallow Test could not have done it without their well-developed EF — their executive function. A second critical ingredient for their success was the motivation to sustain their effort, or grit. For what must have felt like eternity, they kept using their minds and imaginations, shifting their attention, and waiting for the adult to come back without ringing the bell. Two marshmallows — or cookies, or whatever they chose — became their burning goal, strong enough to sustain a heroic effort and make it feel worthwhile. Outside the Surprise Room, the wish list for those we love must surely include a hope that they will discover, stumble into, or create their own burning goals to motivate them to construct the lives they want.

Bruce Springsteen found his goal when he saw himself in a mirror for the first time holding his new guitar. George Ramirez says he found his the first day he spent as a student at KIPP. Mark Owen discovered his in junior high school, when he happened to open *Men in Green Faces,* written by a Navy SEAL, and realized in a flash that that was what he had to become. Dave Levin says that when he started teaching he knew at last why he was put on earth. We all have our story, and we can keep editing it as it unfolds over time — as we look back to figure out what those goals must have been even if we did not know we had them, or look ahead to puzzle out where we seem to be going.

When I was very young, my least favorite uncle was successful in the umbrella-manufacturing business and eager to have me join him. He kept torturing me with questions about what I wanted to be when I grew up, hoping that I would say I wanted to be just like him. For me, that defined exactly what I did not want to be, and instead got me thinking about what I might want to make of myself. Another psychologist, a lifelong colleague and friend who has had one of the most brilliant and successful careers in the history of psychology, attributes his burning goal to his father. During the Great Depression of the 1930s, his father chose to forgo his own ambitions for higher education and achievement to work tirelessly so that his family could survive and thrive. My friend attributes his success to having been driven to make and live the life his father was willing to give up as a tribute in his honor. It became his mission in life.

Self-control skills are essential for pursuing our goals successfully, but it is the goals themselves that give us direction and motivation. They are important determinants of life satisfaction, and those we select early in life have striking effects both on the later goals that we reach and the satisfaction we feel about our lives. No matter how they are formed, the goals that drive our life stories are as important as the EF we need to try to reach them.

Self-control, especially when it is labeled "effortful control," can sound as if it demands a grim commitment to very tough, trying labor — a voluntary entry into a work-driven life of self-denial, of living for the future and missing the pleasures of the moment. An acquaintance told me about a recent dinner he had with friends in Manhattan during which the topic turned to the Marshmallow Test. One of his friends, a novelist who lived in Greenwich Village, was contrasting his own life with that of his brother, a very wealthy and successful investment banker living the pinstriped-suit-with-Hermès-necktie life. The brother had long been married and had children who were all doing well. The writer had published five novels but they had had little impact and few sales. He described himself nevertheless as having a great time, spending his days writing and living the bachelor life at night, going from one short-term relationship to the next. He speculated that his solemn, straight-laced brother probably would have waited forever for his marshmallows, whereas he would have been an early bell ringer.

In fact, the novelist could not have published those five books without a great deal of self-control, and he probably also needs it when trying to maintain his fun relationships while staying uncommitted. Nor did he manage to make it through an elite liberal arts college that emphasizes creative writing without having more than enough self-control to do so. You need EF as much for a creative life in the arts as for a successful life in anything else; it's just the goals that differ. Without EF, the chance to find and pursue your goals is lost. That's what the kids in the

South Bronx faced if they lost in the KIPP lottery. But without compelling goals and drive, EF can leave us competent but aimless.

ALTERNATIVE VIEWS OF WHO WE ARE

Your reactions to the research findings on brain plasticity and the malleability of behavior in this book depend importantly on your own beliefs about how much people can really control and change what they become. There are two conflicting ways to interpret what these findings tell us in the larger context of who we are and what we can be. It is worth using your cool system to think about what the results mean to you before coming to firm conclusions that your hot system has probably already reached.

The answer to the question of whether human nature is, at its core, malleable or fixed has been an enduring concern of not just scientists but, more important, each of us in our everyday lives. Some people see self-control ability, willpower, intelligence, and other characteristics as fixed, unchangeable traits from the very start of life. They read the experimental evidence that executive function and self-control improve after educational interventions and interpret that as short-term effects unlikely to make a long-term difference, just little tricks that don't change inborn traits. These people differ from those who see the evidence as supporting the view that we are open to change and able to alter how we think and behave, that we can craft our own lives rather than being either the winners or the losers in the DNA lottery.

If we allow the evidence to make a difference to our personal theories, the discovery of the plasticity of the brain tells us that human nature is more flexible and open to change than has long been assumed. We do not come into the world with a bundle of fixed, stable traits that determine who we become. We develop in continuous interactions with our social and biological environments. These interactions shape our expectations, the goals and values that drive us, the ways we interpret stimuli and experience, and the life stories we construct.

To reiterate from the nature-nurture discussion (Chapter 7), as Kaufer and Francis point out, "Environments can be as deterministic as we once believed only genes could be, and . . . the genome can be as malleable as we once believed only environments could be." And the basic message of this book has been that there is substantial evidence that we can be active agents who in part control how those interactions play out. That leaves us with a view of human nature in which we potentially have more choice, and more responsibility, than in the purely deterministic scientific views of the past century. Those views attributed the causes of our behavior to the environment, DNA, the unconscious, bad parenting, or evolution, plus chance. The story this book tells acknowledges all these sources as influences. But ultimately, at the end of that causal chain, it is the individual who is the agent of the action and decides when to ring the bell.

When I am asked to summarize the fundamental message from research on self-control, I recall Descartes's famous dictum cogito, ergo sum — "I think, therefore I am." What has been discovered about mind, brain, and self-control lets us move from his proposition to "I think, therefore I can *change* what I am." Because by changing how we think, we can change what we feel, do, and become. If that leads to the question "But can I *really* change?," I reply with what George Kelly said to his therapy clients when they kept asking him if they could get control of their lives. He looked straight into their eyes and said, "Would you like to?"

ACKNOWLEDGMENTS

I am especially indebted to my daughters, Judy Mischel, Rebecca Mischel, and Linda Mischel Eisner, to whom this book is dedicated. When they were children they inspired my research and were its first "subjects"; as adults they generously helped me tell its story. My partner, Michele Tolela Myers, kept me going with wise advice, careful creative editing, and endless support, tolerance, and encouragement. Paul Mischel, my nephew, contributed with his scientific expertise and wisdom, sharp eyes, and caring attention, from start to finish. Ran Hassin was both a cheerleader and a creative editor and adviser at many stages of the manuscript's development. Bert Moore, my student and coinvestigator when the early experiments were done at Stanford and my friend still decades later, read, commented, and reread the manuscript with patience and care. I am grateful to my many colleagues and friends (happily, too many to list individually) for reading and commenting constructively on the manuscript, in whole or in part, often repeatedly.

My agent, John Brockman, believed in this book and helped

make it happen. Tracy Behar, my editor at Little, Brown, edited and reedited every sentence to make it as clear as possible, and Sarah Murphy assisted in this task. Amy Cole, my assistant and right hand, dealt with every detail, small and large, and read and commented on multiple drafts. She worked closely with Brooke Burrows, and together they used my notations to help craft the endnotes.

Finally, I am grateful to the children and families whose contributions and unstinting cooperation, often over the course of many years, created the findings on which most of this book is based. In the same vein, I am deeply indebted to the students and colleagues credited within the manuscript, who became my collaborators and friends, and who conducted the research over my lifetime that enabled this book. The research itself was generously and continuously supported by grants from the National Institute of Mental Health and the National Science Foundation.

NOTES

Introduction

4: **fivefold increase** S. M. Carlson, P. D. Zelazo, and S. Faja, "Executive Function," in *Oxford Handbook of Developmental Psychology*, edited by P. D. Zelazo (New York: Oxford University Press, 2013), 706–743.

4: **"Amy"** Names introduced in quotes are fictitious to protect confidentiality.

5: **functioning in adolescence** W. Mischel, Y. Shoda, and M. L. Rodriguez, "Delay of Gratification in Children," *Science* 244, no. 4907 (1989): 933–938.

6: **the "master aptitude"** D. Goleman, *Emotional Intelligence: The 10th Anniversary Edition* (New York: Bantam Books, 2005), 80–83.

6: **2006, David Brooks** D. Brooks, "Marshmallows and Public Policy," *New York Times*, May 7, 2006.

6: **interview he conducted with President Obama** W. Mischel and D. Brooks "The News from Psychological Science: A Conversation between David Brooks and Walter Mischel," *Perspectives on Psychological Science* 6, no. 6 (2011): 515–520.

6: ***The New Yorker* in a 2009** J. Lehrer, "Don't: The Secret of Self-Control," *The New Yorker*, May 18, 2009.

7: **influencing the curriculum in many schools** See http://www.kipp.org/ and http://www.schoolsthatcan.org/ for examples.

7: **International investment companies** S. Benartzi with R. Lewin, *Save More Tomorrow: Practical Behavioral Finance Solutions to Improve 401(k) Plans* (New York: Penguin Press, 2012).

7: **two closely interacting systems** J. Metcalfe and W. Mischel, "A Hot/Cool System Analysis of Delay of Gratification: Dynamics of Willpower," *Psychological Review* 106, no. 1 (1999): 3–19.

PART I

DELAY ABILITY: ENABLING SELF-CONTROL

1: In Stanford University's Surprise Room

13: **He wrote this summary** T. C. Schelling, *Choice and Consequence: Perspectives of an Errant Economist* (Cambridge, MA: Harvard University Press, 1984), 59.

15: **seeking only immediate satisfaction** L. J. Borstelmann, "Children before Psychology," in *Handbook of Child Psychology: Vol I: History, Theory, and Methods*, 4th ed., edited by P. H. Mussen and W. Kessen (New York: Wiley, 1983), 3–40.

17: **importance of trust** W. Mischel, "Father Absence and Delay of Gratification: Cross-Cultural Comparisons," *Journal of Abnormal and Social Psychology* 63, no. 1 (1961): 116–124; W. Mischel and E. Staub, "Effects of Expectancy on Working and Waiting for Larger Rewards," *Journal of Personality and Social Psychology* 2, no. 5 (1965): 625–633; W. Mischel and J. Grusec, "Waiting for Rewards and Punishments: Effects of Time and Probability on Choice," *Journal of Personality and Social Psychology* 5, no. 1 (1967): 24–31.

21: **predict long-term consequential life outcomes** W. Mischel, *Personality and Assessment* (New York: Wiley, 1968). M. Lewis, "Models of Development," in *Advances in Personality Science*, edited by D. Cervone and W. Mischel (New York: Guilford, 2002), 153–176.

23: **cognitive and social skills at school** W. Mischel, Y. Shoda, and P. K. Peake, "The Nature of Adolescent Competencies Predicted by Preschool Delay of Gratification," *Journal of Personality and Social Psychology* 54, no. 4 (1988): 687–699; W. Mischel, Y. Shoda, and M. L. Rodriguez, "Delay of Gratification in Children," *Science* 244, no. 4907 (1989): 933–938; and Y. Shoda, W. Mischel, and P. K. Peake, "Predicting Adolescent Cognitive and Social Competence from Preschool

Delay of Gratification: Identifying Diagnostic Conditions," *Developmental Psychology* 26, no. 6 (1990): 978–986.

24: earned much better SAT scores Ibid. For links between self-control and intelligence, see also A. L. Duckworth and M. E. Seligman, "Self-Discipline Outdoes IQ in Predicting Academic Performance of Adolescents," *Psychological Science* 16, no. 12 (2005): 939–944; and T. E. Moffitt and others, "A Gradient of Childhood Self-Control Predicts Health, Wealth, and Public Safety," *Proceedings of the National Academy of Sciences* 108, no. 7 (2011): 2693–2698.

24: their scores was 210 points Personal communication from Phil Peake, Smith College, April 9, 2012, and as reported in D. Goleman, *Emotional Intelligence: The 10th Anniversary Edition* (New York: Bantam Books, 2005), 82.

24: Around age twenty-five to thirty O. Ayduk and others, "Regulating the Interpersonal Self: Strategic Self-Regulation for Coping with Rejection Sensitivity," *Journal of Personality and Social Psychology* 79, no. 5 (2000): 776–792.

24: lower body mass index T. R. Schlam and others, "Preschoolers' Delay of Gratification Predicts Their Body Mass 30 Years Later," *Journal of Pediatrics* 162, no. 1 (2013): 90–93.

25: schools of the South Bronx Ayduk, "Regulating the Interpersonal Self."

26: The brain images of these alumni B. J. Casey and others, "Behavioral and Neural Correlates of Delay of Gratification 40 Years Later," *Proceedings of the National Academy of Sciences* 108, no. 36 (2011): 14998–15003.

2: How They Do It

29: "hallucinatory image" S. Freud, "Formulations Regarding the Two Principles of Mental Functioning" in *Collected Papers*, vol. 4, translated by Joane Riviere (New York: Basic Books, 1959).

29: "time binding" D. Rapaport, "Some Metapsychological Considerations Concerning Activity and Passivity," in *The Collected Papers of David Rapaport* (New York: Basic Books, 1967), 530–568.

30: At this age, the children understood W. Mischel and E. B. Ebbesen, "Attention in Delay of Gratification," *Journal of Personality and Social Psychology* 16, no. 2 (1970): 329.

31: the children think some "fun thoughts" W. Mischel, E. B. Ebbesen, and A. R. Zeiss, "Cognitive and Attentional Mechanisms in Delay of

Gratification," *Journal of Personality and Social Psychology* 21, no. 2 (1972): 204–218.

32: slide-projected image W. Mischel and B. Moore, "Effects of Attention to Symbolically Presented Rewards on Self-Control," *Journal of Personality and Social Psychology* 28, no. 2 (1973): 172–179.

33: cued to think about them as if they were real B. Moore, W. Mischel, and A. Zeiss, "Comparative Effects of the Reward Stimulus and Its Cognitive Representation in Voluntary Delay," *Journal of Personality and Social Psychology* 34, no. 3 (1976): 419–424.

34: Daniel Berlyne D. Berlyne, *Conflict, Arousal and Curiosity* (New York: McGraw-Hill, 1980).

34: To test this idea W. Mischel and N. Baker, "Cognitive Appraisals and Transformations in Delay Behavior," *Journal of Personality and Social Psychology* 31, no. 2 (1975): 254.

35: If they thought about fun things Mischel, Ebbesen, and Zeiss, "Cognitive and Attentional Mechanisms in Delay of Gratification."

35: Give nine-year-old children compliments G. Seeman and J. C. Schwarz, "Affective State and Preference for Immediate versus Delayed Reward," *Journal of Research in Personality* 7, no. 4 (1974): 384–394; see also B. S. Moore, A. Clyburn, and B. Underwood, "The Role of Affect in Delay of Gratification," *Child Development* 47, no. 1 (1976): 273–276.

35: what holds for children applies to adults J. R. Gray, "A Bias toward Short-Term Thinking in Threat-Related Negative Emotional States," *Personality and Social Psychology Bulletin* 25, no. 1 (1999): 65–75.

35: less likely to delay gratification when we feel sad E. H. Wertheim and J. C. Schwarz, "Depression, Guilt, and Self-Management of Pleasant and Unpleasant Events," *Journal of Personality and Social Psychology* 45, no. 4 (1983): 884–889.

36: delayed alternatives most often for crayons A. Koriat and M. Nisan. "Delay of Gratification as a Function of Exchange Values and Appetitive Values of the Rewards," *Motivation and Emotion* 2, no. 4 (1978): 375–390.

36: "There is nothing either good or bad" W. Shakespeare, *Hamlet: The New Variorum Edition*, edited by H. H. Furness (Toronto, Ontario: General Publishing Company, 2000), Act II, Scene 2, 245–246.

37: delay gratification increased with age W. Mischel and R. Metzner, "Preference for Delayed Reward as a Function of Age, Intelligence, and Length of Delay Interval," *Journal of Abnormal and Social Psychology* 64, no. 6 (1962): 425–431.

38: thoughts that would make it harder B. T. Yates and W. Mischel, "Young Children's Preferred Attentional Strategies for Delaying Gratification," *Journal of Personality and Social Psychology* 37, no. 2 (1979): 286–300; H. N. Mischel and W. Mischel, "The Development of Children's Knowledge of Self-Control Strategies," *Child Development* 54, no. 3 (1983): 603–619.

39: By age five to six Mischel and Mischel, "The Development of Children's Knowledge of Self-Control Strategies."

40: boys with impulsivity problems M. L. Rodriguez, W. Mischel, and Y. Shoda, "Cognitive Person Variables in the Delay of Gratification of Older Children at Risk," *Journal of Personality and Social Psychology* 57, no. 2 (1989): 358–367.

3: Thinking Hot and Cool

43: hot emotional system For how the hot and cool systems work, see J. Metcalfe and W. Mischel, "A Hot/Cool System Analysis of Delay of Gratification: Dynamics of Willpower," *Psychological Review* 106, no. 1 (1999): 3–19.

44: endure the wait J. A. Gray, *The Psychology of Fear and Stress*, 2nd ed. (New York: McGraw-Hill, 1987); J. LeDoux, *The Emotional Brain* (New York: Simon and Schuster, 1996); J. Metcalfe and W. J. Jacobs, "A 'Hot-System/Cool-System' View of Memory under Stress," *PTSD Research Quarterly* 7, no. 2 (1996): 1–3.

44: Freud called the id S. Freud, "Formulations Regarding the Two Principles of Mental Functioning" in *Collected Papers*, vol. 4, translated by Joane Riviere (New York: Basic Books, 1959).

46: reciprocal relationship Although it is useful to speak and think about "two" systems, they are closely connected brain regions and their neural circuits communicate with each other and interact continuously.

46: The PFC is the most evolved region A. F. Arnsten, "Stress Signaling Pathways That Impair Prefrontal Cortex Structure and Function," *Nature Reviews Neuroscience* 10, no. 6 (2009): 410–422.

47: Age matters H. N. Mischel and W. Mischel, "The Development of Children's Knowledge of Self-Control Strategies," *Child Development* 54, no. 3 (1983): 603–619. For recent work adapting the Marshmallow Test for use at younger ages see P. D. Zelazo and S. M. Carlson, "Hot and Cool Executive Function in Childhood and Adolescence: Development and Plasticity," *Child Development Perspectives* 6, no. 4 (2012): 354–360.

47: In contrast, by age twelve O. Ayduk and others, "Regulating the Interpersonal Self: Strategic Self-Regulation for Coping with Rejection Sensitivity," *Journal of Personality and Social Psychology* 79, no. 5 (2000): 776–792.

48: girls are usually rated higher A. L. Duckworth and M. E. Seligman, "Self-Discipline Gives Girls the Edge: Gender in Self-Discipline, Grades, and Achievement Test Scores," *Journal of Educational Psychology* 98, no. 1 (2006): 198–208.

48: girls are generally more compliant G. Kochanska, K. C. Coy, and K. T. Murray, "The Development of Self-Regulation in the First Four Years of Life," *Child Development* 72, no. 4 (2001): 1091–1111.

48: girls choose delayed rewards more Duckworth and Seligman, "Self-Discipline Gives Girls the Edge."

48: groups studied so far I. W. Silverman, "Gender Differences in Delay of Gratification: A Meta-Analysis," *Sex Roles* 49, nos. 9/10 (2003): 451–463.

49: the choice became hot A. Prencipe and P. D. Zelazo, "Development of Affective Decision Making for Self and Other Evidence for the Integration of First- and Third-Person Perspectives," *Psychological Science* 16, no. 7 (2005): 501–505.

49: Stress can become harmful B. S. McEwen, "Protective and Damaging Effects of Stress Mediators: Central Role of the Brain," *Dialogues in Clinical Neuroscience* 8, no. 4 (2006): 283–297.

50: neuroscientist Amy Arnsten Arnsten, "Stress Signaling Pathways," p. 410; R. M. Sapolsky, "Why Stress Is Bad for Your Brain," *Science* 273, no. 5276 (1996): 749–750.

50: The longer stress persists B. S. McEwen and P. J. Gianaros, "Stress- and Allostasis-Induced Brain Plasticity," *Annual Review of Medicine* 62 (2011): 431–445.

50: Remember Hamlet W. Shakespeare, *Hamlet: The New Variorum Edition*, edited by H. H. Furness (Toronto: General Publishing Company, 2000).

4: The Roots of Self-Control

52: the "Strange Situation" M. D. S. Ainsworth and others, *Patterns of Attachment: A Psychological Study of the Strange Situation* (Hillsdale, NJ: Erlbaum, 1978).

53: "Maternal control" A. Sethi and others, "The Role of Strategic Attention Deployment in Development of Self-Regulation: Predicting

Preschoolers' Delay of Gratification from Mother-Toddler Interactions," *Developmental Psychology* 36, no. 6 (2000): 767.

54: These results underscore G. Kochanska, K. T. Murray, and E. T. Harlan, "Effortful Control in Early Childhood: Continuity and Change, Antecedents, and Implications for Social Development," *Developmental Psychology* 36, no. 2 (2000): 220–232; N. Eisenberg and others, "Contemporaneous and Longitudinal Prediction of Children's Social Functioning from Regulation and Emotionality," *Child Development* 68, no. 4 (1997): 642–664.

54: In the infant's first few months It is the human version of what the rat moms do when they lick and groom (LG) their pups. The rat pups who have high LG mothers perform better on cognitive tasks and display less physiological arousal to acute stress, compared with those who are stuck with low LG mothers (M. J. Meaney, "Maternal Care, Gene Expression, and the Transmission of Individual Differences in Stress Reactivity across Generations," *Annual Review of Neuroscience* 24 (2001): 1161–1192).

54: infants are nurtured C. Harman, M. K. Rothbart, and M. I. Posner, "Distress and Attention Interactions in Early Infancy," *Motivation and Emotion* 21, no. 1 (1997): 27–44; M. I. Posner and M. K. Rothbart, *Educating the Human Brain*, Human Brain Development Series (Washington, DC: APA Books, 2007).

55: infants' stress levels L. A. Sroufe, "Attachment and Development: A Prospective, Longitudinal Study from Birth to Adulthood," *Attachment and Human Development* 7, no. 4 (2005): 349–367; M. Mikulincer and P. R. Shaver, *Attachment Patterns in Adulthood: Structure, Dynamics, and Change* (New York: Guilford Press, 2007).

55: angry-sounding speech while sleeping A. M. Graham, P. A. Fisher, and J. H. Pfeifer, "What Sleeping Babies Hear: A Functional MRI Study of Interparental Conflict and Infants' Emotion Processing," *Psychological Science* 24, no. 5 (2013): 782–789.

55: how their lives unfold Center on the Developing Child at Harvard University, *Building the Brain's "Air Traffic Control" System: How Early Experiences Shape the Development of Executive Function: Working Paper No. 11* (2011).

55: At the neural level Posner and Rothbart, *Educating the Human Brain*.

56: *"mind of their own"* Ibid., 79.

56: follow two simple rules P. D. Zelazo, "The Dimensional Change Card Sort (DCCS): A Method of Assessing Executive Function in Children," *Nature: Protocols* 1, no. 1 (2006): 297–301.

57: **underlying neural circuits** Center on the Developing Child, *Building the Brain's "Air Traffic Control" System.*

57: **The child's experiences** P. D. Zelazo and S. M. Carlson, "Hot and Cool Executive Function in Childhood and Adolescence: Development and Plasticity," *Child Development Perspectives* 6, no. 4 (2012): 354–360.

57: **the intrusiveness with which she inspected, evaluated** P. Roth, *Portnoy's Complaint* (New York: Random House, 1967).

58: **"a man or a mouse?"** Ibid., 16.

59: **They distracted themselves strategically** M. L. Rodriguez and others, "A Contextual Approach to the Development of Self-Regulatory Competencies: The Role of Maternal Unresponsivity and Toddlers' Negative Affect in Stressful Situations," *Social Development* 14, no. 1 (2005): 136–157.

59: **children, aged 12 to 15 months** A. Bernier, S. M. Carlson, and N. Whipple, "From External Regulation to Self-Regulation: Early Parenting Precursors of Young Children's Executive Functioning," *Child Development* 81, no. 1 (2010): 326–339.

60: **parents who overcontrol their toddlers** Sroufe, "Attachment and Development"; and A. A. Hane and N. A. Fox, "Ordinary Variations in Maternal Caregiving Influence Human Infants' Stress Reactivity," *Psychological Science* 17, no. 6 (2006): 550–556.

5: The Best-Laid Plans

61: **"and if I beseech"** S. H. Butcher and A. Lang, *Homer's Odyssey* (London: Macmillan, 1928), 197.

62: **The result was Mr. Clown Box** W. Mischel, "Processes in Delay of Gratification," in *Advances in Experimental Social Psychology,* edited by L. Berkowitz, vol. 7 (New York: Academic Press, 1974), 249–292.

66: **those without this type of plan** W. Mischel and C. J. Patterson, "Substantive and Structural Elements of Effective Plans for Self-Control," *Journal of Personality and Social Psychology* 34, no. 5 (1976): 942–950; C. J. Patterson and W. Mischel, "Effects of Temptation-Inhibiting and Task-Facilitating Plans on Self-Control," *Journal of Personality and Social Psychology* 33, no. 2 (1976): 209–217.

66: **surprisingly powerful *If-Then* plans** For examples of *If-Then* implementation plans, see P. M. Gollwitzer, "Implementation Intentions: Strong Effects of Simple Plans," *American Psychologist* 54, no. 7 (1999): 493–503; P. M. Gollwitzer, C. Gawrilow, and G. Oettingen, "The Power of Planning: Self-Control by Effective Goal-Striving," in *Self*

Control in Society, Mind, and Brain, edited by R. R. Hassin and others (New York: Oxford University Press, 2010), 279–296; G. Stadler, G. Oettingen, and P. Gollwitzer, "Intervention Effects of Information and Self-Regulation on Eating Fruits and Vegetables Over Two Years," *Health Psychology* 29, no. 3 (2010): 274-283.

67: **I will read my textbook** P. M. Gollwitzer, "Goal Achievement: The Role of Intentions," *European Review of Social Psychology* 4, no. 1 (1993): 141–185; P. M. Gollwitzer and V. Brandstätter, "Implementation Intentions and Effective Goal Pursuit," *Journal of Personality and Social Psychology* 73, no. 1 (1997): 186–199.

68: **while your cool system rests** For the concept of two systems, one that "thinks fast" and another that "thinks slow" and is effortful and "lazy," see D. Kahneman, *Thinking, Fast and Slow* (New York: Farrar, Straus and Giroux, 2011).

68: **very difficult laboratory conditions** C. Gawrilow, P. M. Gollwitzer, and G. Oettingen, "If-Then Plans Benefit Executive Functions in Children with ADHD," *Journal of Social and Clinical Psychology* 30, no. 6 (2011): 616–646; and C. Gawrilow and P. M. Gollwitzer, "Implementation Intentions Facilitate Response Inhibition in Children with ADHD," *Cognitive Therapy and Research* 32, no. 2 (2008): 261–280.

6: Idle Grasshoppers and Busy Ants

72: **Consistent with the stereotypes** W. Mischel, "Father Absence and Delay of Gratification: Cross-Cultural Comparisons," *Journal of Abnormal and Social Psychology* 63, no. 1 (1961): 116–124.

73: **experience with a promise maker** Young children's decision making on the marshmallow task is moderated by beliefs about environmental reliability. Ibid.; W. Mischel and E. Staub, "Effects of Expectancy on Working and Waiting for Larger Rewards," *Journal of Personality and Social Psychology* 2, no. 5 (1965): 625–633; W. Mischel and J. C. Masters, "Effects of Probability of Reward Attainment on Responses to Frustration," *Journal of Personality and Social Psychology* 3, no. 4 (1966): 390–396; W. Mischel and J. Grusec, "Waiting for Rewards and Punishments: Effects of Time and Probability on Choice," *Journal of Personality and Social Psychology* 5, no. 1 (1967): 24–31; C. Kidd, H. Palmieri, and R. N. Aslin, "Rational Snacking: Young Children's Decision-Making on the Marshmallow Task Is Moderated by Beliefs about Environmental Reliability," *Cognition* 126, no. 1 (2012): 109–114.

73: Leary was leading the charge D. Lattin, *The Harvard Psychedelic Club: How Timothy Leary, Ram Dass, Huston Smith, and Andrew Weil Killed the Fifties and Ushered In a New Age for America* (New York: HarperCollins, 2011).

73: schools in the Boston area W. Mischel and C. Gilligan, "Delay of Gratification, Motivation for the Prohibited Gratification, and Resistance to Temptation," *Journal of Abnormal and Social Psychology* 69, no. 4 (1964): 411–417.

75: chosen to wait for larger This was an early demonstration that such choice preferences can predict important behavior like gaining weight, excessive risk taking, drug use, etc. Researchers now often use such choices as a shortcut measure when the Marshmallow Test cannot be used.

76: McClure and his team See S. M. McClure and others, "Separate Neural Systems Value Immediate and Delayed Monetary Rewards," *Science* 306, no. 5695 (2004): 503–507.

77: headed by Elke Weber and Bernd Figner B. Figner and others, "Lateral Prefrontal Cortex and Self-Control in Intertemporal Choice," *Nature Neuroscience* 13, no. 5 (2010): 538–539.

77: wants what it wants immediately For an alternative interpretation of these results see J. W. Kable and P. W. Glimcher, "An 'As Soon as Possible' Effect in Human Intertemporal Decision Making: Behavioral Evidence and Neural Mechanisms," *Journal of Neurophysiology* 103, no. 5 (2010): 2513–2531.

78: "idiosyncrasies of human preferences" McClure, "Separate Neural Systems," 506.

78: temptation in the particular situation E. Tsukayama and A. L. Duckworth, "Domain-Specific Temporal Discounting and Temptation," *Judgment and Decision Making* 5, no. 2 (2010): 72–82.

78: "I can resist everything except temptation" O. Wilde, *Lady Windermere's Fan: A Play about a Good Woman*, Act I (1892). For research on the same point, see E. Tsukayama, A. L. Duckworth, and B. Kim, "Resisting Everything Except Temptation: Evidence and an Explanation for Domain-Specific Impulsivity," *European Journal of Personality* 26, no. 3 (2011): 318–334.

7: Is It Prewired? The New Genetics

79: "the Irish intellect" J. D. Watson with A. Berry, *DNA: The Secret of Life* (New York: Knopf Doubleday Publishing Group, 2003), 361.

81: **American psychology into the 1950s** B. F. Skinner, *Science and Human Behavior* (New York: Macmillan, 1953).

81: **The newborn's slate** S. Pinker, *The Blank Slate: The Modern Denial of Human Nature* (New York: Penguin, 2003).

82: **babies are born accountants** N. Angier, "Insights from the Youngest Minds," *New York Times*, May 3, 2012; F. Xu, E. S. Spelke, and S. Goddard, "Number Sense in Human Infants," *Developmental Science* 8, no. 1 (2005): 88–101.

82: **Babies enter the world** M. K. Rothbart, L. K. Ellis, and M. I. Posner, "Temperament and Self-Regulation," in *Handbook of Self-Regulation: Research, Theory, and Applications*, edited by K. D. Vohs and R. F. Baumeister (New York: Guilford, 2011), 441–460.

83: **discuss how their baby's temperament** A. H. Buss and R. Plomin, *Temperament: Early Developing Personality Traits* (Hillsdale, NJ: Erlbaum, 1984); D. Watson and L. A. Clark, "The PANAS-X: Manual for the Positive and Negative Affect Schedule — Expanded Form," University of Iowa, Iowa Research Online (1999); and M. K. Rothbart and S. A. Ahadi, "Temperament and the Development of Personality," *Journal of Abnormal Psychology* 103, no. 1 (1994): 55–66.

84: **a reasonable estimate from twin research** S. H. Losoya and others, "Origins of Familial Similarity in Parenting: A Study of Twins and Adoptive Siblings," *Developmental Psychology* 33, no. 6 (1997): 1012; R. Plomin, "The Role of Inheritance in Behavior," *Science* 248, no. 4952 (1990): 183–188.

84: **parse nature and nurture** W. Mischel, Y. Shoda, and O. Ayduk, *Introduction to Personality: Toward an Integrative Science of the Person*, 8th ed. (New York: Wiley, 2008).

84: **nature and nurture are not easily separated** D. Kaufer and D. Francis, "Nurture, Nature, and the Stress That Is Life," in *Future Science: Cutting-Edge Essays from the New Generation of Scientists*, edited by M. Brockman (New York: Oxford University Press, 2011), 56–71.

85: **well-sharpened pencil** Mischel, Shoda, and Ayduk, *Introduction to Personality*.

87: **Given these discoveries** F. A. Champagne and R. Mashoodh, "Genes in Context: Gene-Environment Interplay and the Origins of Individual Differences in Behavior," *Current Directions in Psychological Science* 18, no. 3 (2009): 127–131.

87: **mothers exposed to violence** K. M. Radtke and others, "Transgenerational Impact of Intimate Partner Violence on Methylation in the

Promoter of the Glucocorticoid Receptor," *Translational Psychiatry* 1, no. 7 (2011): e21.

87: **Stress in childhood** D. D. Francis and others, "Maternal Care, Gene Expression, and the Development of Individual Differences in Stress Reactivity," *Annals of the New York Academy of Sciences* 896, no. 1 (1999): 66–84.

87: **non-genomic characteristics of our cells are inherited** Ibid.; I. C. Weaver and others, "Epigenetic Programming by Maternal Behavior," *Nature Neuroscience* 7, no. 8 (2004): 847–854.

89: **active agent in her own development** L. A. Schmidt and N. A. Fox, "Individual Differences in Childhood Shyness: Origins, Malleability, and Developmental Course," in *Advances in Personality Science*, edited by D. Cervone and W. Mischel (New York: Guilford, 2002), 83–105.

89: **genetically brave mice placed with shy mothers** D. D. Francis and others, "Epigenetic Sources of Behavioral Differences in Mice," *Nature Neuroscience* 6, no. 5 (2003): 445–446.

90: **"maze-dull" or "maze-bright"** R. M. Cooper and J. P. Zubek, "Effects of Enriched and Restricted Early Environments on the Learning Ability of Bright and Dull Rats," *Canadian Journal of Psychology/Revue Canadienne de Psychologie* 12, no. 3 (1958): 159–164.

90: **high LG mothers benefited greatly** M. J. Meaney, "Maternal Care, Gene Expression, and the Transmission of Individual Differences in Stress Reactivity across Generations," *Annual Review of Neuroscience* 24, no. 1 (2001): 1161–1192.

91: **upward trend in IQ scores** J. R. Flynn, "The Mean IQ of Americans: Massive Gains 1932 to 1978," *Psychological Bulletin* 95, no. 1 (1984): 29–51; J. R. Flynn, "Massive IQ Gains in 14 Nations: What IQ Tests Really Measure," *Psychological Bulletin* 101, no. 2 (1987): 171–191.

91: **"A predisposition does not a predetermination make"** Watson and Berry, *DNA: The Secret of Life*, 391.

91: **a study in New Zealand** A. Caspi and others, "Influence of Life Stress on Depression: Moderation by a Polymorphism in the 5-HTT Gene," *Science* 301, no. 5631 (2003): 386–389.

92: **even political beliefs** Mischel, Shoda, and Ayduk, *Introduction to Personality*.

92: **"inverting implicit assumptions"** Kaufer and Francis, "Nurture, Nature, and the Stress That Is Life," 63.

PART II

FROM MARSHMALLOWS IN PRE-K TO MONEY IN 401(K)

96: well-intentioned policemen B. K. Payne, "Weapon Bias: Split-Second Decisions and Unintended Stereotyping," *Current Directions in Psychological Science* 15, no. 6 (2006): 287–291.

8: The Engine of Success: "I Think I Can!"

102: George Ramirez Source for material in this section: personal interview with George Ramirez, March 14, 2013, at KIPP Academy Middle School, South Bronx; G. Ramirez, unpublished autobiography, March 2013; and G. Ramirez, "Changed by the Bell," *Yale Herald*, February 17, 2012.

105: "I probably worked harder" D. Remnick, "*New Yorker* Profiles: 'We Are Alive' — Bruce Springsteen at Sixty-Two," *The New Yorker*, July 30, 2012, 56.

107: executive function (EF) EF is sometimes called executive control or EC.

107: three features of EF E. T. Berkman, E. B. Falk, and M. D. Lieberman, "Interactive Effects of Three Core Goal Pursuit Processes on Brain Control Systems: Goal Maintenance, Performance Monitoring, and Response Inhibition," *PLoS ONE* 7, no. 6 (2012): e40334.

108: Cognitive scientists can now see P. D. Zelazo and S. M. Carlson, "Hot and Cool Executive Function in Childhood and Adolescence: Development and Plasticity," *Child Development Perspectives* 6, no. 4 (2012): 354–360; B. J. Casey and others, "Behavioral and Neural Correlates of Delay of Gratification 40 Years Later," *Proceedings of the National Academy of Sciences* 108, no. 36 (2011): 14998–15003; and M. I. Posner and M. K. Rothbart, *Educating the Human Brain*, Human Brain Development Series (Washington, DC: APA Books, 2007).

108: EF allows planning C. Blair, "School Readiness: Integrating Cognition and Emotion in a Neurobiological Conceptualization of Children's Functioning at School Entry," *American Psychologist* 57, no. 2 (2002): 111–127; and R. A. Barkley, "The Executive Functions and Self-Regulation: An Evolutionary Neuropsychological Perspective," *Neuropsychology Review* 11, no. 1 (2001): 1–29.

108: It is not a surprise K. L. Bierman and others, "Executive Functions and School Readiness Intervention: Impact, Moderation, and Mediation in the Head Start REDI Program," *Development and Psychopathology* 20, no. 3 (2008): 821–843; and M. M. McClelland and others, "Links between Behavioral Regulation and Preschoolers' Literacy,

Vocabulary, and Math Skills," *Developmental Psychology* 43, no. 3 (2007): 947–959.

108: circuits involved in EF are closely interconnected Posner and Rothbart, *Educating the Human Brain.*

109: Children who lack EF N. Eisenberg and others, "The Relations of Emotionality and Regulation to Children's Anger-Related Reactions," *Child Development* 65, no. 1 (1994): 109–128; A. L. Hill and others, "Profiles of Externalizing Behavior Problems for Boys and Girls across Preschool: The Roles of Emotion Regulation and Inattention," *Developmental Psychology* 42, no. 5 (2006): 913–928; and G. Kochanska, K. Murray, and K. C. Coy, "Inhibitory Control as a Contributor to Conscience in Childhood: From Toddler to Early School Age," *Child Development* 68, no. 2 (1997): 263–277.

109: These skills help kids not only delay M. L. Rodriguez, W. Mischel, and Y. Shoda, "Cognitive Person Variables in the Delay of Gratification of Older Children at Risk," *Journal of Personality and Social Psychology* 57, no. 2 (1989): 358–367; and O. Ayduk, W. Mischel, and G. Downey, "Attentional Mechanisms Linking Rejection to Hostile Reactivity: The Role of 'Hot' versus 'Cool' Focus," *Psychological Science* 13, no. 5 (2002): 443–448.

109: they can be interpersonally cool E. Tsukayama, A. L. Duckworth, and B. E. Kim, "Domain-Specific Impulsivity in School-Age Children," *Developmental Science* 16, no. 6 (2013): 879–893.

110: development of flexible and adaptive self-control S. M. Carlson and R. F. White, "Executive Function, Pretend Play, and Imagination," in *The Oxford Handbook of the Development of Imagination*, edited by M. Taylor (New York: Oxford University Press, 2013).

110: "theory of mind" S. M. Carlson and L. J. Moses, "Individual Differences in Inhibitory Control and Children's Theory of Mind," *Child Development* 72, no. 4 (2001): 1032–1053.

111: "By feeling, not by thinking" Giacomo Rizzolatti quoted in S. Blakeslee, "Cells That Read Minds," *New York Times*, January 10, 2006.

111: negative effects of stress S. E. Taylor and A. L. Stanton, "Coping Resources, Coping Processes, and Mental Health," *Annual Review of Clinical Psychology* 3 (2007): 377–401.

112: As Taylor and her colleagues reported in 2011 S. Saphire-Bernstein and others, "Oxytocin Receptor Gene (OXTR) Is Related to Psychological Resources," *Proceedings of the National Academy of Sciences* 108, no. 37 (2011): 15118; and B. S. McEwen, "Protective and Damaging

Effects of Stress Mediators: Central Role of the Brain," *Dialogues in Clinical Neuroscience* 8, no. 4 (2006): 283–297.

112: you can be an active agent A. Bandura, *Self-Efficacy: The Exercise of Control* (New York: Freeman, 1997); and A. Bandura, "Toward a Psychology of Human Agency," *Perspectives on Psychological Science* 1, no. 2 (2006): 164–180.

113: people's personal theories C. Dweck, *Mindset: The New Psychology of Success* (New York: Random House, 2006).

114: "I suck in math" Ibid., 57.

115: "I think I can!" W. Piper, *The Little Engine That Could* (New York: Penguin, 1930).

115: "internal versus external control" W. Mischel, R. Zeiss, and A. Zeiss, "Internal-External Control and Persistence: Validation and Implications of the Stanford Preschool Internal-External Scale," *Journal of Personality and Social Psychology* 29, no. 2 (1974): 265–278.

116: The child's self-perception Bandura, "Toward a Psychology of Human Agency."

116: The children's sense of efficacy and agency M. R. Lepper, D. Greene, and R. E. Nisbett, "Undermining Children's Intrinsic Interest with Extrinsic Reward: A Test of the 'Overjustification' Hypothesis," *Journal of Personality and Social Psychology* 28, no. 1 (1973): 129–137; and E. L. Deci, R. Koestner, and R. M. Ryan, "A Meta-Analytic Review of Experiments Examining the Effects of Extrinsic Rewards on Intrinsic Motivation," *Psychological Bulletin* 125, no. 6 (1999): 627–668.

117: optimists cope more effectively S. E. Taylor and D. A. Armor, "Positive Illusions and Coping with Adversity," *Journal of Personality* 64, no. 4 (1996): 873–898; and Saphire-Bernstein and others, "Oxytocin Receptor Gene (OXTR) Is Related to Psychological Resources." See also C. S. Carver, M. F. Scheier, and S. C. Segerstrom, "Optimism," *Clinical Psychology Review* 30, no. 7 (2010): 879–889.

117: coronary bypass surgery M. E. Scheier, J. K. Weintraub, and C. S. Carver, "Coping with Stress: Divergent Strategies of Optimists and Pessimists," *Journal of Personality and Social Psychology* 51, no. 6 (1986): 1257–1264.

117: "I really hate" W. T. Cox and others, "Stereotypes, Prejudice, and Depression: The Integrated Perspective," *Perspectives on Psychological Science* 7, no. 5 (2012): 427–449.

117: They attribute the bad things L. Y. Abramson, M. E. Seligman, and J. D. Teasdale, "Learned Helplessness in Humans: Critique and Reformulation," *Journal of Abnormal Psychology* 87, no. 1 (1978): 49–74.

118: pessimistic explanatory style C. Peterson, M. E. Seligman, and G. E. Valliant, "Pessimistic Explanatory Style Is a Risk Factor for Physical Illness: A Thirty-Five-Year Longitudinal Study," *Journal of Personality and Social Psychology* 55, no. 1 (1988): 23–27.

118: players were all outstanding enough C. Peterson and M. E. Seligman "Explanatory Style and Illness," *Journal of Personality* 55, no. 2 (1987): 237–265.

119: optimists deal with failure constructively Interview with Seligman reported in D. Goleman, "Research Affirms Power of Positive Thinking," *New York Times*, February 3, 1987. See also M. E. Scheier and C. S. Carver, "Dispositional Optimism and Physical Well-Being: The Influence of Generalized Outcome Expectancies on Health," *Journal of Personality* 55, no. 2 (1987): 169–210; and Carver, Scheier, and Segerstrom, "Optimism."

119: explanatory style tells you who gives up Quoted in D. Goleman, *Emotional Intelligence*, 10th Anniversary Edition (New York: Bantam Books, 2005), 88–89.

9: Your Future Self

124: All the world's a stage J. P. Kimble, ed., *Shakespeare's As You Like It: A Comedy* (London: S. Gosnell, Printer, 1810), Act II, Scene 7, 139–166.

125: travel through time H. Ersner-Hershfield and others, "Don't Stop Thinking about Tomorrow: Individual Differences in Future Self-Continuity Account for Saving," *Judgment and Decision Making* 4, no. 4 (2009): 280–286.

126: the stranger pattern This discussion draws extensively on "The Face Tool" section in S. Benartzi with R. Lewin, *Save More Tomorrow: Practical Behavioral Finance Solutions to Improve 401(k) Plans* (New York: Penguin Press, 2012).

126: Hal Hershfield, now at New York University H. Ersner-Hershfield, G. E. Wimmer, and B. Knutson, "Saving for the Future Self: Neural Measures of Future Self-Continuity Predict Temporal Discounting," *Social Cognitive and Affective Neuroscience* 4, no. 1 (2009): 85–92.

126: The same group of researchers Ersner-Hershfield and others, "Don't Stop Thinking about Tomorrow."

127: representations of their retirement-age selves H. E. Hershfield and others, "Increasing Saving Behavior through Age-Progressed Renderings of the Future Self," *Journal of Marketing Research: Special Issue* 48, SPL (2011): 23–37.

129: in 401(k) retirement plans Benartzi, *Save More Tomorrow,* 142–158; Hershfield and others, "Increasing Saving Behavior"; S. M. McClure and others, "Separate Neural Systems Value Immediate and Delayed Monetary Rewards," *Science* 306, no. 5695 (2004): 503–507.

129: unethical but profitable business decisions H. E. Hershfield, T. R. Cohen, and L. Thompson, "Short Horizons and Tempting Situations: Lack of Continuity to Our Future Selves Leads to Unethical Decision Making and Behavior," *Organizational Behavior and Human Decision Processes* 117, no. 2 (2012): 298–310.

10: Beyond the Here and Now

132: Psychologists Yaacov Trope and Nira Liberman Y. Trope and N. Liberman, "Construal Level Theory," in *Handbook of Theories of Social Psychology,* vol. 1, edited by P. A. M. Van Lange and others (New York: Sage Publications, 2012), 118–134; N. Liberman and Y. Trope, "The Psychology of Transcending the Here and Now," *Science* 322, no. 5905 (2008): 1201–1205.

133: why people make decisions D. T. Gilbert and T. D. Wilson, "Prospection: Experiencing the Future," *Science* 317, no. 5843 (2007): 1351–1354.

133: The psychological immune system D. T. Gilbert and J. E. Ebert, "Decisions and Revisions: The Affective Forecasting of Changeable Outcomes," *Journal of Personality and Social Psychology* 82, no. 4 (2002): 503–514; D. Gilbert, *Stumbling on Happiness* (New York: Knopf, 2006); and D. Kahneman and J. Snell, "Predicting a Changing Taste: Do People Know What They Will Like?," *Journal of Behavioral Decision Making* 5, no. 3 (1992): 187–200.

133: imagine yourself doing it in the present D. I. Tamir and J. P. Mitchell, "The Default Network Distinguishes Construals of Proximal versus Distal Events," *Journal of Cognitive Neuroscience* 23, no. 10 (2011): 2945–2955.

133: Such high-level, abstract thinking What Metcalfe and Mischel ("A Hot/Cool System Analysis of Delay of Gratification: Dynamics of Willpower," *Psychological Review* 106, no. 1 [1999]: 3–19) call the hot system overlaps with what other researchers call the default system

(Tamir and Mitchell, "The Default Network") or the visceral system (G. Loewenstein, "Out of Control: Visceral Influences on Behavior," *Organizational Behavior and Human Decision Processes* 65, no. 3 [1996]: 272–292) or System 1 (D. Kahneman, *Thinking, Fast and Slow* [New York: Farrar, Straus and Giroux, 2011]).

134: reduces the automatic preference K. Fujita and others, "Construal Levels and Self-Control," *Journal of Personality and Social Psychology* 90, no. 3 (2006): 351–367.

134: Recall that when preschoolers Ibid.; W. Mischel and B. Moore, "Effects of Attention to Symbolically Presented Rewards on Self-Control," *Journal of Personality and Social Psychology* 28, no. 2 (1973): 172–179; W. Mischel and N. Baker, "Cognitive Appraisals and Transformations in Delay Behavior," *Journal of Personality and Social Psychology* 31, no. 2 (1975): 254.

134: Kevin Ochsner and his team H. Kober and others, "Prefrontal-Striatal Pathway Underlies Cognitive Regulation of Craving," *Proceedings of the National Academy of Sciences* 107, no. 33 (2010): 14811–14816.

134: cognitively regulated their appetitive impulses J. A. Silvers and others, "Neural Links between the Ability to Delay Gratification and Regulation of Craving in Childhood." Society for Neuroscience Annual Meeting, San Diego, CA, 2013.

136: stifle the craving For regulation of craving by cognitive strategies in cigarette smokers, see Kober, "Prefrontal-Striatal Pathway"; R. E. Bliss and others, "The Influence of Situation and Coping on Relapse Crisis Outcomes after Smoking Cessation," *Journal of Consulting and Clinical Psychology* 57, no. 3 (1989): 443–449; S. Shiffman and others, "First Lapses to Smoking: Within-Subjects Analysis of Real-Time Reports," *Journal of Consulting and Clinical Psychology* 64, no. 2 (1996): 366–379.

137: penetrated my hot system As George Loewenstein ("Out of Control") noted, doctors generally smoke less than most people, but the difference is greatest among those who regularly deal with images of the smoke-blackened lungs of their diseased patients.

138: aversive counterconditioning W. Mischel, Y. Shoda, and O. Ayduk, *Introduction to Personality: Toward an Integrative Science of the Person*, 8th ed. (New York: Wiley, 2008).

141: improve informed consent for DNA testing This work was done in collaboration also with Yuichi Shoda.

142: "pre-living" scenarios Y. Shoda and others, "Psychological Interventions and Genetic Testing: Facilitating Informed Decisions about BRCA1/2

Cancer Susceptibility," *Journal of Clinical Psychology in Medical Settings* 5, no. 1 (1998): 3–17. See also S. J. Curry and K. M. Emmons, "Theoretical Models for Predicting and Improving Compliance with Breast Cancer Screening," *Annals of Behavioral Medicine* 16, no. 4 (1994): 302–316.

144: "You, however, are not convinced that all is well" S. M. Miller, "Monitoring and Blunting: Validation of a Questionnaire to Assess Styles of Information Seeking under Threat," *Journal of Personality and Social Psychology* 52, no. 2 (1987): 345–353.

144: categorized as "monitors" S. M. Miller and C. E. Mangan, "Interacting Effects of Information and Coping Style in Adapting to Gynecologic Stress: Should the Doctor Tell All?," *Journal of Personality and Social Psychology* 45, no. 1 (1983): 223–236.

145: if there is nothing you can do to reduce the stress Miller and Mangan, "Interacting Effects of Information and Coping Style"; and S. M. Miller, "Monitoring versus Blunting Styles of Coping with Cancer Influence the Information Patients Want and Need about Their Disease: Implications for Cancer Screening and Management," *Cancer* 76, no. 2 (1995): 167–177.

11: Protecting the Hurt Self: Self-Distancing

147: psychological distancing and cognitive reappraisal A. Luerssen and O. Ayduk, "The Role of Emotion and Emotion Regulation in the Ability to Delay Gratification," in *Handbook of Emotion Regulation*, 2nd ed., edited by J. Gross (2014); and E. Kross and O. Ayduk, "Facilitating Adaptive Emotional Analysis: Distinguishing Distanced-Analysis of Depressive Experiences from Immersed-Analysis and Distraction," *Personality and Social Psychology Bulletin* 34, no. 7 (2008): 924–938.

148: many others get worse S. Nolen-Hoeksema, "The Role of Rumination in Depressive Disorders and Mixed Anxiety/Depressive Symptoms," *Journal of Abnormal Psychology* 109, no. 3 (2000): 504–511; S. Nolen-Hoeksema, B. E. Wisco, and S. Lyubomirsky, "Rethinking Rumination," *Perspectives on Psychological Science* 3, no. 5 (2008): 400–424.

149: experiments on self-distancing E. Kross, O. Ayduk, and W. Mischel, "When Asking 'Why' Does Not Hurt: Distinguishing Rumination from Reflective Processing of Negative Emotions," *Psychological Science* 16, no. 9 (2005): 709–715.

151: In a 2010 experiment O. Ayduk and E. Kross, "From a Distance: Implications of Spontaneous Self-Distancing for Adaptive

Self-Reflection," *Journal of Personality and Social Psychology* 98, no. 5 (2010): 809–829.

151: rumination: elevated blood pressure O. Ayduk and E. Kross, "Enhancing the Pace of Recovery: Self-Distanced Analysis of Negative Experiences Reduces Blood Pressure Reactivity," *Psychological Science* 19, no. 3 (2008): 229–231.

152: large twenty-one-day daily diary study Ayduk and Kross, "From a Distance," study 3.

154: reappraising intensely negative stimuli J. J. Gross and O. P. John, "Individual Differences in Two Emotion Regulation Processes: Implications for Affect, Relationships, and Well-Being," *Journal of Personality and Social Psychology* 85, no. 2 (2003): 348–362; and K. N. Ochsner and J. J. Gross, "Cognitive Emotion Regulation Insights from Social Cognitive and Affective Neuroscience," *Current Directions in Psychological Science* 17, no. 2 (2008): 153–158.

154: linked in earlier research K. A. Dodge, "Social-Cognitive Mechanisms in the Development of Conduct Disorder and Depression," *Annual Review of Psychology* 44, no. 1 (1993): 559–584; K. L. Bierman and others, "School Outcomes of Aggressive-Disruptive Children: Prediction from Kindergarten Risk Factors and Impact of the Fast Track Prevention Program," *Aggressive Behavior* 39, no. 2 (2013): 114–130.

154: boys and girls were cued E. Kross and others, "The Effect of Self-Distancing on Adaptive versus Maladaptive Self-Reflection in Children," *Emotion-APA* 11, no. 5 (2011): 1032–1039.

156: emotional pain really do hurt E. Kross and others, "Social Rejection Shares Somatosensory Representations with Physical Pain," *Proceedings of the National Academy of Sciences* 108, no. 15 (2011): 6270–6275.

156: Naomi Eisenberger and her colleagues N. I. Eisenberger, M. D. Lieberman, and K. D. Williams, "Does Rejection Hurt? An fMRI Study of Social Exclusion," *Science* 302, no. 5643 (2003): 290–292.

157: This antidote E. Selcuk and others, "Mental Representations of Attachment Figures Facilitate Recovery Following Upsetting Autobiographical Memory Recall," *Journal of Personality and Social Psychology* 103, no. 2 (2012): 362–378.

12: Cooling Painful Emotions

159: destructive effects of high RS R. Romero-Canyas and others, "Rejection Sensitivity and the Rejection-Hostility Link in Romantic

Relationships," *Journal of Personality* 78, no. 1 (2010): 119–148; and G. Downey and others, "The Self-Fulfilling Prophecy in Close Relationships: Rejection Sensitivity and Rejection by Romantic Partners," *Journal of Personality and Social Psychology* 75, no. 2 (1998): 545–560.

160: RS children are more easily victimized V. Purdie and G. Downey, "Rejection Sensitivity and Adolescent Girls' Vulnerability to Relationship-Centered Difficulties," *Child Maltreatment* 5, no. 4 (2000): 338–349.

160: making depression more likely O. Ayduk, W. Mischel, and G. Downey, "Attentional Mechanisms Linking Rejection to Hostile Reactivity: The Role of 'Hot' versus 'Cool' Focus," *Psychological Science* 13, no. 5 (2002): 443–448; O. Ayduk, G. Downey, and M. Kim, "Rejection Sensitivity and Depressive Symptoms in Women," *Personality and Social Psychology Bulletin* 27, no. 7 (2001): 868–877.

160: dorsal anterior cingulate cortex and anterior insula G. Bush, P. Luu, and M. I. Posner, "Cognitive and Emotional Influences in Anterior Cingulate Cortex," *Trends in Cognitive Sciences* 4, no. 6 (2000): 215–222. See also G. M. Slavich and others, "Neural Sensitivity to Social Rejection Is Associated with Inflammatory Responses to Social Stress," *Proceedings of the National Academy of Sciences* 107, no. 33 (2010): 14817–14822.

161: long-term inflammation R. M. Sapolsky, L. M. Romero, and A. U. Munck, "How Do Glucocorticoids Influence Stress Responses? Integrating Permissive, Suppressive, Stimulatory, and Preparative Actions," *Endocrine Reviews* 21, no. 1 (2000): 55–89.

161: lower self-esteem, lower self-worth O. Ayduk and others, "Regulating the Interpersonal Self: Strategic Self-Regulation for Coping with Rejection Sensitivity," *Journal of Personality and Social Psychology* 79, no. 5 (2000): 776–792.

162: coped as well in their lives O. Ayduk and others, "Rejection Sensitivity and Executive Control: Joint Predictors of Borderline Personality Features," *Journal of Research in Personality* 42, no. 1 (2008): 151–168.

165: High RS youngsters were less accepted O. Ayduk and others, "Regulating the Interpersonal Self."

166: "When I get criticism, I write it down" For the benefits of writing about emotional experiences see J. W. Pennebaker, *Opening Up: The Healing Power of Expressing Emotion* (New York: Guilford Press, 1997), and J. W. Pennebaker, "Writing about Emotional Experiences as a Therapeutic Process," *Psychological Science* 8, no. 3 (1997): 162–166.

167: predicted their body mass index T. R. Schlam and others, "Preschoolers' Delay of Gratification Predicts Their Body Mass 30 Years Later," *Journal of Pediatrics* 162, no. 1 (2012): 91.

168: thousand children born in Dunedin T. E. Moffitt and others, "A Gradient of Childhood Self-Control Predicts Health, Wealth, and Public Safety," *Proceedings of the National Academy of Sciences* 108, no. 7 (2011): 2693–2698.

13: The Psychological Immune System

169: "psychological immune system" Daniel Gilbert discusses both the psychological and the biological immune systems in *Stumbling on Happiness* (New York: Knopf, 2006), 162. For how the psychological immune system also leads to poor predictions of future happiness, see D. T. Gilbert and T. D. Wilson, "Prospection: Experiencing the Future," *Science* 317, no. 5843 (2007): 1351–1354; and D. T. Gilbert and others, "Immune Neglect: A Source of Durability Bias in Affective Forecasting," *Journal of Personality and Social Psychology* 75, no. 3 (1998): 617–638.

171: better than they rate their peers S. E. Taylor and D. A. Armor, "Positive Illusions and Coping with Adversity," *Journal of Personality* 64, no. 4 (1996): 873–898; and S. E. Taylor and P. M. Gollwitzer, "Effects of Mindset on Positive Illusions," *Journal of Personality and Social Psychology* 69, no. 2 (1995): 213–226.

171: "routinely sum to more than 100 percent" D. G. Myers, "Self-Serving Bias," in *This Will Make You Smarter: New Scientific Concepts to Improve Your Thinking*, edited by J. Brockman (New York: Harper Perennial, 2012), 37–38.

172: High self-enhancers have a healthier HPA axis profile S. E. Taylor and others, "Are Self-Enhancing Cognitions Associated with Healthy or Unhealthy Biological Profiles?," *Journal of Personality and Social Psychology* 85, no. 4 (2003): 605–615.

173: self-affirming mental states S. E. Taylor and others, "Psychological Resources, Positive Illusions, and Health," *American Psychologist* 55, no. 1 (2000): 99–109.

173: perceive themselves more accurately D. A. Armor and S. E. Taylor, "When Predictions Fail: The Dilemma of Unrealistic Optimism," in *Heuristics and Biases: The Psychology of Intuitive Judgment*, edited by T. Gilovich, D. Griffin, and D. Kahneman (New York: Cambridge University Press, 2002), 334–347; and S. E. Taylor and J. D. Brown, "Illusion and Well-Being: A Social Psychological Perspective on Mental Health," *Psychological Bulletin* 103, no. 2 (1988): 193–210.

173: somewhat illusory, glow M. D. Alicke, "Global Self-Evaluation as Determined by the Desirability and Controllability of Trait Adjectives," *Journal of Personality and Social Psychology* 49, no. 6 (1985): 1621–1630; and G. W. Brown and others, "Social Support, Self-Esteem and Depression," *Psychological Medicine* 16, no. 4 (1986): 813–831.

173: as Daniel Gilbert points out Gilbert, *Stumbling on Happiness,* 162.

174: Aaron Beck A. T. Beck and others, *Cognitive Therapy of Depression* (New York: Guilford Press, 1979).

174: clinically depressed patients evaluate their performance P. M. Lewinsohn and others, "Social Competence and Depression: The Role of Illusory Self-Perceptions," *Journal of Abnormal Psychology* 89, no. 2 (1980): 203–212.

176: inflation in self-evaluation L. B. Alloy and L. Y. Abramson, "Judgment of Contingency in Depressed and Nondepressed Students: Sadder but Wiser?," *Journal of Experimental Psychology: General* 108, no. 4 (1979): 441–485.

176: happy and sad feelings impacted performance J. Wright and W. Mischel, "Influence of Affect on Cognitive Social Learning Person Variables," *Journal of Personality and Social Psychology* 43, no. 5 (1982): 901–914; see also A. M. Isen and others, "Affect, Accessibility of Material in Memory, and Behavior: A Cognitive Loop?," *Journal of Personality and Social Psychology* 36, no. 1 (1978): 1–12.

177: They evaluated themselves as more intelligent To learn about regulating and cooling anxiety and other negative emotions, see J. Gross, "Emotion Regulation: Taking Stock and Moving Forward," *Emotion* 13, no. 3 (2013): 359–365; and K. N. Ochsner and others, "Rethinking Feelings: An fMRI Study of the Cognitive Regulation of Emotion," *Journal of Cognitive Neuroscience* 14, no. 8 (2002): 1215–1229.

178: friendships that were just as long lasting S. E. Taylor and others, "Portrait of the Self-Enhancer: Well Adjusted and Well Liked or Maladjusted and Friendless?," *Journal of Personality and Social Psychology* 84, no. 1 (2003): 165–176.

179: maybe his goal was to impress himself S. M. Carlson and L. J. Moses, "Individual Differences in Inhibitory Control and Children's Theory of Mind," *Child Development* 72, no. 4 (2001): 1032–1053.

179: individual strengths and self-regard E. Diener and M. E. Seligman, "Very Happy People," *Psychological Science* 13, no. 1 (2002): 81–84; E. L. Deci and R. M. Ryan, eds., *Handbook of Self-Determination Research* (Rochester, NY: University of Rochester Press, 2002).

180: leads to overconfidence D. Kahneman, *Thinking, Fast and Slow* (New York: Farrar, Straus and Giroux, 2011).

181: The scandal of General Petraeus S. Shane and S. G. Stolberg, "A Brilliant Career with a Meteoric Rise and an Abrupt Fall," *New York Times*, November 10, 2012.

182: it was simulated and analyzed by Maria Konnikova M. Konnikova, *The Limits of Self-Control: Self-Control, Illusory Control, and Risky Financial Decision Making*, PhD dissertation, Columbia University, 2013.

183: "An optimistic bias plays a role" Kahneman, *Thinking, Fast and Slow*, 256.

183: 1,100 new inventions T. Astebro, "The Return to Independent Invention: Evidence of Unrealistic Optimism, Risk Seeking or Skewness Loving?," *Economic Journal* 113, no. 484 (2003): 226–239; and T. Astebro and S. Elhedhli, "The Effectiveness of Simple Decision Heuristics: Forecasting Commercial Success for Early-Stage Ventures," *Management Science* 52, no. 3 (2006): 395–409.

184: "completely certain" of their diagnosis Reported in Kahneman, *Thinking, Fast and Slow*, 263, based on E. S. Berner and M. L. Graber, "Overconfidence as a Cause of Diagnostic Error in Medicine," *American Journal of Medicine* 121, no. 5 (2008): S2–S23.

184: Early in my career W. Mischel, *Personality and Assessment* (New York: Wiley, 1968).

184: The weight of the patients' folders Mischel, *Personality and Assessment*; and J. J. Lasky and others, "Post-Hospital Adjustment as Predicted by Psychiatric Patients and by Their Staff," *Journal of Consulting Psychology* 23, no. 3 (1959): 213–218.

185: the candidates' simple self-reports W. Mischel, "Predicting the Success of Peace Corps Volunteers in Nigeria," *Journal of Personality and Social Psychology* 1, no. 5 (1965): 510–517.

186: a similar lack of validity Kahneman, *Thinking, Fast and Slow*.

186: attempting to walk on hot coals C. Pogash, "A Self-Improvement Quest That Led to Burned Feet," *New York Times*, July 22, 2012.

14: When Smart People Act Stupid

187: a person who lies and cheats R. V. Burton, "Generality of Honesty Reconsidered," *Psychological Review* 70, no. 6 (1963): 481–499.

188: Judge Wachtler was famously revered J. M. Caher, *King of the Mountain: The Rise, Fall, and Redemption of Chief Judge Sol Wachtler* (Amherst, NY: Prometheus Books, 1998).

188: Tiger Woods J. Surowiecki, "Branded a Cheat," *The New Yorker*, December 21, 2009.

190: "Only the little people pay taxes" D. Gilson, "Only Little People Pay Taxes," *Mother Jones*, April 18, 2011.

191: We make these judgments easily W. Mischel, *Personality and Assessment* (New York: Wiley, 1968); W. Mischel, Y. Shoda, and O. Ayduk, *Introduction to Personality: Toward an Integrative Science of the Person*, 8th ed. (New York: Wiley, 2008).

192: The assumption that people are broadly consistent D. T. Gilbert and P. S. Malone, "The Correspondence Bias," *Psychological Bulletin* 117, no. 1 (1995), 21–38; M. D. Lieberman and others, "Reflexion and Reflection: A Social Cognitive Neuroscience Approach to Attributional Inference," *Advances in Experimental Social Psychology* 34 (2002): 199–249; Mischel, *Personality and Assessment*.

192: failed to support the core trait assumption H. Hartshorne, M. A. May, and J. B. Maller, *Studies in the Nature of Character, II Studies in Service and Self-Control* (New York: Macmillan, 1929); Mischel, *Personality and Assessment*; W. Mischel, "Toward an Integrative Science of the Person (Prefatory Chapter)," *Annual Review of Psychology* 55 (2004): 1–22; T. Newcomb, "The Consistency of Certain Extrovert-Introvert Behavior Patterns in Fifty-One Problem Boys," *Teachers College Record* 31, no. 3 (1929): 263–265; W. Mischel and P. K. Peake, "Beyond Déjà Vu in the Search for Cross-Situational Consistency," *Psychological Review* 89, no. 6 (1982): 730–755.

193: In 1968, I undertook a comprehensive review Mischel, *Personality and Assessment*.

193: failed to demonstrate the consistency of behavior J. Block, "Millennial Contrarianism: The Five-Factor Approach to Personality Description 5 Years Later," *Journal of Research in Personality* 35, no. 1 (2001): 98–107; W. Mischel, "Toward a Cognitive Social Learning Reconceptualization of Personality," *Psychological Review* 80, no. 4 (1973): 252–283;

W. Mischel, "From *Personality and Assessment* (1968) to Personality Science," *Journal of Research in Personality* 43, no. 2 (2009): 282–290.

193: While the debate continued I. Van Mechelen, "A Royal Road to Understanding the Mechanisms Underlying Person-in-Context Behavior," *Journal of Research in Personality* 43, no. 2 (2009): 179–186; and V. Zayas and Y. Shoda, "Three Decades after the Personality Paradox: Understanding Situations," *Journal of Research in Personality* 43, no. 2 (2009): 280–281.

194: predict specific behavior Kahneman, *Thinking, Fast and Slow*; Mischel, *Personality and Assessment*; Van Mechelen, "A Royal Road to Understanding."

194: my research team and I did find consistency J. C. Wright and W. Mischel, "A Conditional Approach to Dispositional Constructs: The Local Predictability of Social Behavior," *Journal of Personality and Social Psychology* 53, no. 6 (1987): 1159–1177; and W. Mischel and Y. Shoda, "A Cognitive-Affective System Theory of Personality: Reconceptualizing Situations, Dispositions, Dynamics, and Invariance in Personality Structure," *Psychological Review* 102, no. 2 (1995): 246–268.

15: *If-Then* Signatures of Personality

196: Descriptions by the counselors J. C. Wright and W. Mischel, "Conditional Hedges and the Intuitive Psychology of Traits," *Journal of Personality and Social Psychology* 55, no. 3 (1988): 454–469.

197: classic and intuitively compelling conception See W. Mischel, *Personality and Assessment* (New York: Wiley, 1968); and W. Mischel, "Toward an Integrative Science of the Person (Prefatory Chapter)," *Annual Review of Psychology* 55 (2004): 1–22.

200: *If-Then* patterns tend to be fairly stable Key findings and methods are in Y. Shoda, W. Mischel, and J. C. Wright, "Intraindividual Stability in the Organization and Patterning of Behavior: Incorporating Psychological Situations into the Idiographic Analysis of Personality," *Journal of Personality and Social Psychology* 67, no. 4 (1994): 674–687; W. Mischel and Y. Shoda, "A Cognitive-Affective System Theory of Personality: Reconceptualizing Situations, Dispositions, Dynamics, and Invariance in Personality Structure," *Psychological Review* 102, no. 2 (1995): 246–268.

200: guide treatment and educational plans A. L. Zakriski, J. C. Wright, and M. K. Underwood, "Gender Similarities and Differences in Children's Social Behavior: Finding Personality in Contextualized Patterns of Adaptation," *Journal of Personality and Social Psychology* 88, no. 5

(2006): 844–855; and R. E. Smith and others, "Behavioral Signatures at the Ballpark: Intraindividual Consistency of Adults' Situation-Behavior Patterns and Their Interpersonal Consequences," *Journal of Research in Personality* 43, no. 2 (2009): 187–195.

200: Since the Wediko research M. A. Fournier, D. S. Moskowitz, and D. C. Zuroff, "Integrating Dispositions, Signatures, and the Interpersonal Domain," *Journal of Personality and Social Psychology* 94, no. 3 (2008): 531–545; I. Van Mechelen, "A Royal Road to Understanding the Mechanisms Underlying Person-in-Context Behavior," *Journal of Research in Personality* 43, no. 2 (2009): 179–186; and O. Ayduk and others, "Verbal Intelligence and Self-Regulatory Competencies: Joint Predictors of Boys' Aggression," *Journal of Research in Personality* 41, no. 2 (2007): 374–388.

200: behavioral signature of personality Mischel and Shoda, "A Cognitive-Affective System Theory of Personality."

201: What we found for aggression at Wediko W. Mischel and P. K. Peake, "Beyond Déjà Vu in the Search for Cross-Situational Consistency," *Psychological Review* 89, no. 6 (1982): 730–755.

202: illusions of consistency Mischel and Shoda, "A Cognitive-Affective System Theory of Personality."

202: stability of our *If-Then* patterns Mischel and Peake, "Beyond Déjà Vu in the Search for Cross-Situational Consistency."

203: how you decide to evaluate his overall behavior W. Mischel, "Continuity and Change in Personality," *American Psychologist* 24, no. 11 (1969): 1012–1018.

204: carefully structured daily diaries Y. Shoda and others, "Cognitive-Affective Processing System Analysis of Intra-Individual Dynamics in Collaborative Therapeutic Assessment: Translating Basic Theory and Research into Clinical Applications," *Journal of Personality* 81, no. 6 (2013): 554–568.

205: did not use a cooling strategy These relationships were influenced importantly by the child's intelligence. See Ayduk, "Verbal Intelligence and Self-Regulatory Competencies."

16: The Paralyzed Will

206: "The Angel of the Bridge" J. Cheever, "The Angel of the Bridge," *The New Yorker*, October 21, 1961.

207: Within the hot memories of his amygdala J. LeDoux, *The Emotional Brain* (New York: Simon and Schuster, 1996); J. LeDoux, "Parallel

Memories: Putting Emotions Back into the Brain," in *The Mind: Leading Scientists Explore the Brain, Memory, Personality, and Happiness*, edited by J. Brockman (New York: HarperCollins, 2011), 31–47.

207: The unfortunate dogs For a discussion of this kind of "classical conditioning" see W. Mischel, Y. Shoda, and O. Ayduk, *Introduction to Personality: Toward an Integrative Science of the Person*, 8th ed. (New York: Wiley, 2008), Chapter 10.

209: "If a response antagonistic to anxiety" J. Wolpe, *Reciprocal Inhibition Therapy* (Stanford, CA: Stanford University Press, 1958), 71.

210: Cheever's story was a preview A. Bandura, *Principles of Behavior Modification* (New York: Holt, Rinehart and Winston, 1969); G. L. Paul, *Insight vs. Desensitization in Psychotherapy* (Stanford, CA: Stanford University Press, 1966); and A. T. Beck and others, *Cognitive Therapy of Depression* (New York: Guilford Press, 1979).

210: just petted the dog Bandura, *Principles of Behavior Modification.*

210: Bandura's research showed Ibid.; A. Bandura, J. E. Grusec, and F. L. Menlove, "Vicarious Extinction of Avoidance Behavior," *Journal of Personality and Social Psychology* 5, no. 1 (1967): 16–23; and A. Bandura and F. L. Menlove, "Factors Determining Vicarious Extinction of Avoidance Behavior through Symbolic Modeling," *Journal of Personality and Social Psychology* 8, no. 2 (1968): 99–108.

210: "guided mastery experiences" L. Williams, "Guided Mastery Treatment of Agoraphobia: Beyond Stimulus Exposure," in *Progress in Behavior Modification*, vol. 26, edited by M. Hersen, R. M. Eisler, and P. M. Miller (Newbury Park, CA: Sage, 1990), 89–121.

211: "The changes endured" A. Bandura, "Albert Bandura," in *A History of Psychology in Autobiography*, vol. 9, edited by G. Lindzey and W. M. Runyan (Washington, DC: American Psychological Association, 2006), 62–63.

211: Gordon Paul assigned college students Paul, *Insight vs. Desensitization in Psychotherapy;* G. L. Paul, "Insight versus Desensitization in Psychotherapy Two Years after Termination," *Journal of Consulting Psychology* 31, no. 4 (1967): 333–348.

17: Will Fatigue

216: strength model of self-control M. Muraven, D. M. Tice, and R. F. Baumeister, "Self-Control as Limited Resource: Regulatory Depletion

Patterns," *Journal of Personality and Social Psychology* 74, no. 3 (1998): 774–789.

217: Radish Experiment R. F. Baumeister and others, "Ego Depletion: Is the Active Self a Limited Resource?," *Journal of Personality and Social Psychology* 74, no. 5 (1998): 1252–1265.

217: no matter which act of self-control R. F. Baumeister and J. Tierney, *Willpower: Rediscovering the Greatest Human Strength* (New York: Penguin Press, 2011).

218: not caused by the reasons M. Inzlicht and B. J. Schmeichel, "What Is Ego Depletion? Toward a Mechanistic Revision of the Resource Model of Self-Control," *Perspectives on Psychological Science* 7, no. 5 (2012): 450–463.

218: As motivation to exert self-control increases M. Muraven and E. Slessareva, "Mechanisms of Self-Control Failure: Motivation and Limited Resources," *Personality and Social Psychology Bulletin* 29, no. 7 (2003): 894–906.

219: Their later performance was not impaired C. Martijn and others, "Getting a Grip on Ourselves: Challenging Expectancies about Loss of Energy after Self-Control," *Social Cognition* 20, no. 6 (2002): 441–460.

219: after a strenuous experience V. Job, C. S. Dweck, and G. M. Walton, "Ego Depletion — Is It All in Your Head? Implicit Theories about Willpower Affect Self-Regulation," *Psychological Science* 21, no. 11 (2010): 1686–1693.

220: These findings underscore the importance See also D. C. Molden and others, "Motivational versus Metabolic Effects of Carbohydrates on Self-Control," *Psychological Science* 23, no. 10 (2012): 1137–1144.

221: French parenting P. Druckerman, *Bringing Up Bébé: One American Mother Discovers the Wisdom of French Parenting* (New York: Penguin Press, 2012).

221: Chinese American mother A. Chua, *Battle Hymn of the Tiger Mother* (London: Bloomsbury, 2011).

221: A dozen years earlier J. R. Harris, *The Nurture Assumption: Why Kids Turn Out the Way They Do* (London: Bloomsbury, 1998).

222: adult models influence A. Bandura, "Vicarious Processes: A Case of No-Trial Learning," in *Advances in Experimental Social Psychology*, vol. 2, edited by L. Berkowitz (New York: Academic Press, 1965), 1–55.

223: model's attributes and self-reward behavior W. Mischel and R. M. Liebert, "Effects of Discrepancies between Observed and Imposed Reward Criteria on Their Acquisition and Transmission," *Journal of*

Personality and Social Psychology 3, no. 1 (1966): 45–53; W. Mischel and R. M. Liebert, "The Role of Power in the Adoption of Self-Reward Patterns," *Child Development* 38, no. 3 (1967): 673–683.

225: learned from a model who was lenient with herself The impact of models depends on characteristics like their warmth, nurturance, and power. See J. Grusec and W. Mischel, "Model's Characteristics as Determinants of Social Learning," *Journal of Personality and Social Psychology* 4, no. 2 (1966): 211–215; and W. Mischel and J. Grusec, "Determinants of the Rehearsal and Transmission of Neutral and Aversive Behaviors," *Journal of Personality and Social Psychology* 3, no. 2 (1966): 197–205.

225: This study suggests Models also powerfully influence children's willingness to choose larger delayed rewards rather than smaller immediate rewards. See A. Bandura and W. Mischel, "Modification of Self-Imposed Delay of Reward Through Exposure to Live and Symbolic Models," *Journal of Personality and Social Psychology* 2, no. 5 (1965): 698–705.

225: Mark Owen M. Owen with K. Maurer, *No Easy Day: The First-Hand Account of the Mission That Killed Osama bin Laden* (New York: Dutton, 2012).

226: "these men every day" Ibid., author's note, XI.

PART III

FROM LAB TO LIFE

18: Marshmallows and Public Policy

234: led me to a research career W. Mischel, "Walter Mischel," in *A History of Psychology in Autobiography*, vol. 9, edited by G. E. Lindzey and W. M. Runyan (Washington, DC: American Psychological Association, 2007), 229–267.

234–35: The unexpected finding B. S. McEwen and P. J. Gianaros, "Stress- and Allostasis-Induced Brain Plasticity," *Annual Review of Medicine* 62 (2011): 431–445; Center on the Developing Child at Harvard University, *Building the Brain's "Air Traffic Control" System: How Early Experiences Shape the Development of Executive Function: Working Paper No. 11* (2011); and M. I. Posner and M. K. Rothbart, *Educating the Human Brain*, Human Brain Development Series (Washington, DC: APA Books, 2007).

235: training and genetics jointly influence M. R. Rueda and others, "Training, Maturation, and Genetic Influences on the Development of

Executive Attention," *Proceedings of the National Academy of Sciences* 102, no. 41 (2005): 14931–14936.

237: Given the importance of EF A. Diamond and others, "Preschool Program Improves Cognitive Control," *Science* 318, no. 5855 (2007): 1387–1388; and N. R. Riggs and others, "The Mediational Role of Neurocognition in the Behavioral Outcomes of a Social-Emotional Prevention Program in Elementary School Students: Effects of the PATHS Curriculum," *Prevention Science* 7, no. 1 (2006): 91–102.

238: At age 11 to 12 years C. Gawrilow, P. M. Gollwitzer, and G. Oettingen, "If-Then Plans Benefit Executive Functions in Children with ADHD," *Journal of Social and Clinical Psychology* 30, no. 6 (2011); and C. Gawrilow and others, "Mental Contrasting with Implementation Intentions Enhances Self-Regulation of Goal Pursuit in Schoolchildren at Risk for ADHD," *Motivation and Emotion* 37, no. 1 (2013): 134–145.

238: not only improved their working memory T. Klingberg and others, "Computerized Training of Working Memory in Children with ADHD — a Randomized, Controlled Trial," *Journal of the American Academy of Child and Adolescent Psychiatry* 44, no. 2 (2005): 177–186.

238: Simple meditation and mindfulness Y. Y. Tang and others, "Short-Term Meditation Training Improves Attention and Self-Regulation," *Proceedings of the National Academy of Sciences* 104, no. 43 (2007): 17152–17156; A. P. Jha, J. Krompinger, and M. J. Baime, "Mindfulness Training Modifies Subsystems of Attention," *Cognitive, Affective, & Behavioral Neuroscience* 7, no. 2 (2007): 109–119. See also M. K. Rothbart and others, "Enhancing Self-Regulation in School and Clinic," in *Minnesota Symposia on Child Psychology: Meeting the Challenge of Translational Research in Child Psychology*, vol. 35, edited by M. R. Gunner and D. Cicchetti (Hoboken, NJ: Wiley, 2009), 115–158.

239: Mindfulness training M. D. Mrazek and others, "Mindfulness Training Improves Working Memory Capacity and GRE Performance While Reducing Mind Wandering," *Psychological Science* 24, no. 5 (2013): 776–781.

239: Two of the most notable McEwen and Gianaros, "Stress- and Allostasis-Induced Brain Plasticity."

239: "that these capacities can be improved" Center on the Developing Child, *Building the Brain's "Air Traffic Control" System*, 12.

240: as one pundit put it D. Brooks, "When Families Fail," *New York Times*, February 12, 2013.

241: ***Sesame Street* sets out a challenge** "Sesame Workshop"®, "Sesame Street"®, and associated characters, trademarks, and design elements are owned and licensed by Sesame Workshop. © 2013 Sesame Workshop. All rights reserved.

244: The *Sesame Street* education researchers S. Fisch and R. Truglio, eds., "The Early Window Project: *Sesame Street* Prepares Children for School," in *"G" Is for Growing: Thirty Years of Research on Sesame Street* (Mahwah, NJ: Erlbaum, 2001), 97–114.

244: the "biology of disadvantage" N. E. Adler and J. Stewart, eds., *The Biology of Disadvantage: Socioeconomic Status and Health* (Boston, MA: Wiley-Blackwell, 2010).

245: transform public education Robin Hood Excellence Program, supported by Paul Tudor-Jones, and Michael Druckman's Schools That Can are other examples of the many diverse efforts currently being pursued.

245: impoverished backgrounds Defined by qualifying for the free or reduced-cost lunch program.

247: "Madeline," age ten Personal interview with KIPP student, March 14, 2013, at KIPP Academy Middle School, South Bronx, NY.

250: This compares with a rate These data are from Mischel interviews with Dave Levin, February 22, 2013, and with Mitch Brenner, April 17, 2013.

250: "If you want kids to learn" Personal communication from Dave Levin at KIPP to Mischel on December 26, 2013.

252: almost verbatim one of the qualities Y. Shoda, W. Mischel, and P. K. Peake, "Predicting Adolescent Cognitive and Social Competence from Preschool Delay of Gratification: Identifying Diagnostic Conditions," *Developmental Psychology* 26, no. 6 (1990): 978–986.

252: the earlier preschool years In some states this reflects the fact that preschool education is not funded by the state.

19: Applying Core Strategies

254: humans and animals G. Ainslie and R. J. Herrnstein, "Preference Reversal and Delayed Reinforcement," *Animal Learning and Behavior* 9, no. 4 (1981): 476–482.

254: simple mathematical model D. Laibson, "Golden Eggs and Hyperbolic Discounting," *Quarterly Journal of Economics* 112, no. 2 (1997): 443–478.

258: ***If-Then* implementation plans has helped** P. M. Gollwitzer and G. Oettingen, "Goal Pursuit," in *The Oxford Handbook of Human Motivation,* edited by R. M. Ryan (New York: Oxford University Press, 2012), 208–231.

259: maintaining the change over time R. W. Jeffery and others, "Long-Term Maintenance of Weight Loss: Current Status," *Health Psychology* 19, no. 1S (2000): 5–16.

260: Precommitment strategies M. J. Crockett and others, "Restricting Temptations: Neural Mechanisms of Precommitment," *Neuron* 79, no. 2 (2013): 391–401.

262: strategy you want to try See for example D. Ariely and K. Wertenbroch, "Procrastination, Deadlines, and Performance: Self-Control by Precommitment," *Psychological Science* 13, no. 3 (2002): 219–224.

262: when the default was enrollment D. Laibson, "Psychological and Economic Voices in the Policy Debate," presentation at Psychological Science and Behavioral Economics in the Service of Public Policy, the White House, Washington, DC, May 22, 2013. See also R. H. Thaler and C. R. Sunstein, *Nudge: Improving Decisions about Health, Wealth, and Happiness* (New York: Penguin, 2008).

265: change in perspective alters how the experience is appraised E. Kross and others, "Asking Why from a Distance: Its Cognitive and Emotional Consequences for People with Major Depressive Disorder," *Journal of Abnormal Psychology* 121, no. 3 (2012): 559–569; and E. Kross and O. Ayduk, "Making Meaning out of Negative Experiences by Self-Distancing," *Current Directions in Psychological Science* 20, no. 3 (2011): 187–191.

266: help with the toughest struggles B. A. Alford and A. T. Beck, *The Integrative Power of Cognitive Therapy* (New York: Guilford Press, 1998); and A. T. Beck and others, *Cognitive Therapy of Depression* (New York: Guilford Press, 1979).

266: angry voices while they are sleeping A. M. Graham, P. A. Fisher, and J. H. Pfeifer, "What Sleeping Babies Hear: A Functional MRI Study of Interparental Conflict and Infants' Emotion Processing," *Psychological Science* 24, no. 5 (2013): 782–789.

268: "He was biting" Quotes from personal communication with "Elizabeth" on August 27, 2013.

271: we'd better be careful to keep our promises L. Michaelson and others, "Delaying Gratification Depends on Social Trust," *Frontiers in Psychology* 4 (2013): 355; W. Mischel, "Processes in Delay of Gratification," in *Advances in Experimental Social Psychology*, edited by L. Berkowitz, vol. 7 (New York: Academic Press, 1974), 249–292.

271: they were left to play by themselves A. Bandura, D. Ross, and S. A. Ross, "Transmission of Aggression through Imitation of Aggressive

Models," *Journal of Abnormal and Social Psychology* 63, no. 3 (1961): 575–582.

20: Human Nature

273: **"written in our genes"** Radiolab: http://www.radiolab.org/story/96056-your-future-marshmallow/.

273: **A main lesson from modern science** P. D. Zelazo and W. A. Cunningham, "Executive Function: Mechanisms Underlying Emotion Regulation," in *Handbook of Emotion Regulation,* edited by J. J. Gross (New York: Guilford Press, 2007), 135–158; and Center on the Developing Child at Harvard University, *Building the Brain's "Air Traffic Control" System: How Early Experiences Shape the Development of Executive Function: Working Paper No. 11* (2011).

274: **Springsteen found his goal** P. A. Carlin, *Bruce* (New York: Touchstone, 2012), 24.

275: **important determinants of life satisfaction** Originally published in W. G. Bowen and D. Bok, *The Shape of the River: Long-Term Consequences of Considering Race in College and University Admissions* (Princeton, NJ: Princeton University Press, 1998); and C. Nickerson, N. Schwarz, and E. Diener, "Financial Aspirations, Financial Success, and Overall Life Satisfaction: Who? And How?," *Journal of Happiness Studies* 8, no. 4 (2007): 467–515. For a summary of the essential findings see D. Kahneman, *Thinking, Fast and Slow* (New York: Farrar, Straus and Giroux, 2011), 401–402.

277: **human nature is, at its core** W. Mischel, "Continuity and Change in Personality," *American Psychologist* 24, no. 11 (1969): 1012–1018; and W. Mischel, "Toward an Integrative Science of the Person (Prefatory Chapter)," *Annual Review of Psychology* 55 (2004): 1–22.

278: **the life stories that we construct** C. M. Morf and W. Mischel, "The Self as a Psycho-Social Dynamic Processing System: Toward a Converging Science of Selfhood," in *Handbook of Self and Identity,* 2nd ed., edited by M. Leary and J. Tangney (New York: Guilford, 2012), 21–49.

278: **"environments could be"** D. Kaufer and D. Francis, "Nurture, Nature, and the Stress That Is Life," in *Future Science: Cutting-Edge Essays from the New Generation of Scientists,* edited by M. Brockman (New York: Oxford University Press, 2011), 63.

278: **Descartes's famous dictum** R. Descartes, *Principles of Philosophy,* Part I, article 7 (1644).

INDEX

Note: Italic page numbers refer to illustrations.

ABOUT THE AUTHOR

Walter Mischel was born in Vienna, Austria. After the takeover by the Nazis in 1938, he escaped to the United States as a young child with his family. He has a BA from New York University, an MA in clinical psychology from the City College of New York, and a PhD in clinical psychology from Ohio State University. He taught briefly at the University of Colorado, at Harvard from 1958 to 1962, and then served for 21 years as a professor and chair at Stanford University. Since 1983 he has been at Columbia University, where he holds the Robert Johnston Niven chair as professor of humane letters in psychology.

Mischel is internationally known as the creator of the Marshmallow Test, one of the most famous and important experiments in the history of psychology. His experiments, begun with preschool children at Stanford University's Bing Nursery School in the late 1960s, opened the way for the modern scientific analysis of the cognitive mechanisms that enable delay of gratification and self-control beginning early in life. His work has yielded surprisingly strong predictions for consequential health and well-being outcomes over the life course, while also illuminating the underlying processes and cognitive skills essential for "willpower."

Mischel's work has transformed thinking in his field and shaped much of the modern agenda in the study of individual differences in social behavior and self-control. He was elected to the National Academy of Sciences in 2004 and became a Fellow, American Academy of Arts and Sciences in 1991. He received the Distinguished Scientific Contribution Award from the American Psychological Association in 1982; the Distinguished Scientist Award, Society of Experimental Social Psychologists in 2000; and the Distinguished Scientist Award, American Psychological Association, Division of Clinical Psychology in 1978. His many prizes include the Ludwig Wittgenstein Prize (2012) and the Grawemeyer Award for Psychology (2011).